W9-DJE-129

WITHDRAWN

St. Louis Community College

Library

5801 Wilson Avenue
St. Louis, Missouri 63110

ROBOTICS
IN
PRACTICE

Management and applications
of industrial robots

Associate authors:
Dennis Lock and Kenneth Willis

Robotics
in
Practice

Management and applications
of industrial robots

Joseph F. Engelberger

With a Foreword by Isaac Asimov

A Division of
AMERICAN MANAGEMENT
ASSOCIATIONS

Library of Congress Cataloging in Publication Data

Engelberger, Joseph F
  Robotics in practice.

  Bibliography: p.
  Includes index.
  1. Robots, Industrial.
  I. Title.
  T59.4.E53 1980   629.8'92   80-66866

  ISBN 0-8144-5645-6

Printed in the United States of America.

Seventh printing 1984

Contents

List of illustrations and color plates   ix
Foreword by Isaac Asimov   xiii
Author's preface   xv

PART 1 Fundamentals and management

1. Robot use in manufacturing   3

    Evolution of industrial robots, 3
    Near relations of the robot, 7
    Robot cost *versus* human labor, 9
    Die casting — an early success story for industrial robots, 12
    Robots *versus* special-purpose automation, 15

2. Robot anatomy   19

    Robot classification, 19
    Arm geometry, 30
    Drive systems, 33
    Dynamic performance and accuracy, 35

3. End effectors: hands, grippers, pickups and tools   41

    Methods of grasping, 41
    Mechanical grippers, 42
    Vacuum systems, 49
    Magnetic pickups, 51
    Tools, 55

4. Matching robots to the workplace   59

    Part orientation, 59
    Interlocks and sequence control, 61
    Workplace layout, 67

5. Reliability, maintenance and safety 75

Environmental factors in robot systems, 75
Designing robots for industrial environments, 78
Reliability targets, 82
Theoretical reliability assessment, 83
Maintenance needs and economics, 85
Safety levels and precautions, 89

6. Organizing to support robotics 93

Example of manufacturer's training system, 93
How General Electric built an in-house capability, 95
Work force acceptance of robots, 97

7. Robot economics 110

Checklist of economic factors: costs and benefits, 101
Project appraisal by the payback method, 104
Return on investment evaluation, 107
Areas of cost exposure, 109

8. Sociological impact of robots 111

Quality of working life, 111
Attitudes to robots, 112
Effect on employment, 115

9. Future capabilities 117

Future attributes of robots, 117
Commentary on future attributes, 120
Priorities in attribute development, 125
Interaction with other technologies, 128
Future applications, 133

PART II Application studies

10. Die casting applications 141

Outline of die casting operation, 141
Robots in die casting, 145
Further considerations for robot die casting, 155

11. Spot welding applications 159

Outline of spot welding operation, 159
Robots in spot welding, 163
Planning a robot spot welding line, 164

12. Arc welding applications   171

    Arc welding process, 171
    Robots in arc welding, 174
    Programming the robot, 176
    Choice of robots for arc welding, 177
    Case example of arc welding robot, 178
    Flame cutting: a related application, 179

13. Investment casting applications   181

    The investment casting process, 181
    Mold making by robot, 184
    Basic programs for robot mold making, 186
    Case example at Pratt & Whitney, 187

14. Forging applications   189

    Forging processes, 189
    The working environment of the forging process, 192
    Robots in forging, 193

15. Press work applications   197

    Press operations, 197
    Current applications of robots in the press shop, 199
    Outlook for further robot handling of press work, 203

16. Spray painting applications   207

    Paint behavior and the technique of painting, 207
    The spray painting environment, 208
    Automation in the paint spraying industry, 209
    Robots in paint spraying, 210
    Outlook for robot painting in the automotive industry, 212
    Benefits analysis of robot painting, 214

17. Plastic molding applications   217

    Plastic molding processes, 217
    Opportunities for robot applications, 220
    Current robot use in plastic molding, 220

18. Applications in foundry practice   225

    The casting process, 225
    Robots in the foundry, 227
    Applying robots to the fettling operation, 228

19. Machine tool loading applications   233

    Development of automation in the machine shop, 233
    Robot applications to machine tools, 235
    Robot attributes for machine tool applications, 243

20. Heat treatment applications   247
   Heat treatment processes, 247
   Robots in heat treatment, 249

21. Applications for deburring metal parts   253
   Demands of the deburring operation, 253
   Robot requirements for deburring, 254

22. Palletizing applications   257
   Robot use to achieve optimal pallet loading, 257
   Depalletizing by robot, 260

23. Applications in brick manufacture   263
   The brick manufacture process, 263
   The robot contribution to brickmaking, 265

24. Applications in glass manufacture   269
   Outline of glass manufacturing process, 269
   Robot handling of sheet glass, 271
   Robot handling of fragile glass products, 273

Appendix: List of principal robot manufacturers   277

Bibliography   279

Index   285

List of illustrations and color plates

Figure no.

1.1	Comparison of human and robot characteristics	6
1.2	Robot characteristics — extended specification	8
1.3	Some near relations of robots	10
1.4	Labor cost escalation in the U.S. automotive industry	11
1.5	Payback evaluation of robot costs	12
2.1	Schematic arrangement of a typical limited sequence robot	21
2.2	Analog servo system	24
2.3	Task for a playback robot with point-to-point control	25
2.4	Typical articulations of a playback robot with point-to-point control	26
2.5	Teach pendant for instructing playback robot with point-to-point control	27
2.6	Robot arm configurations	31
2.7	Typical wrist articulations	32
2.8	Diagram of robot arm performance	36
2.9	Graph of robot arm performance	37
2.10	Elements of a single articulation servo system	38
2.11	Typical velocity traces for long and short arm motions	39
3.1	Example showing calculation of grasping force	43
3.2	Examples of mechanical grippers	45
3.3	Some typical vacuum pickup systems	52
3.4	Typical electro magnet pickup for use with flat surfaces	54
3.5	Examples of tools fastened to robot wrists	55
4.1	Sequence control example: the workpieces	64
4.2	Sequence control example: workpiece feed positions	65
4.3	Sequence control example: equipment layout	66
4.4	Work comes to robot	69
4.5	Work travels past robot — diagram of tracking and control system	70
4.6	Work travels past robot — examples of tracking windows	72
4.7	Robot travels to work — track mounted robot serving eleven machine tools	73

Figure no.

4.8	Robot travels to work — overhead robot serving eight NC lathes	74
4.9	Robot travels to work — diagram of overhead robot system portrayed in Figure 4.8	74
5.1	Hazards in the industrial environment	76
5.2	Hazardous situation: robot services die casting machine	78
5.3	Hazardous situation: robot transferring billets in and out of rotary furnace	79
5.4	Hazardous situation: robot re-entering press bed	80
5.5	Hazardous situation: robot protected against machining chips	81
5.6	Hazardous situation: robot subjected to sparks, oil leaks and water spray on spot welding line	82
5.7	Reliability of electronic/electrical elements used in Unimate 2000 Series design	84
5.8	Unimate system reliability estimate	85
5.9	Reliability control points in the Unimate life cycle	87
5.10	The Unimate line at General Motors' Lordstown plant	88
6.1	Outline of seminar on general applications of industrial robots, as conducted by Unimation, Inc.	94
6.2	Work force acceptance checklist	98
7.1	Simple payback example	105
7.2	Complex payback example	106
7.3	Example of return on investment calculation	108
7.4	Return on investment graph	109
9.1	Robot qualities already commercially available	117
9.2	Robot qualities sought for the future	118
9.3	Disciplines useful to the robotics game	119
9.4	Compliance device for mating parts	126
9.5	Diagram of laboratory setup for evaluating robot sensory perception and manipulator dynamics	127
9.6	Advanced technologies contributing to productivity improvement	128
9.7	Hierarchical control system for robot installation	132
9.8	Robot designed to human size	137
9.9	Introduction of robot arms into conventional indexing assembly line	137
9.10	Human size robot — the Unimate 500	138
9.11	Attributes of VAL computer language for addressing PUMA robot	138
10.1	Elements of the hot-chamber die casting machine	142
10.2	Elements of the cold-chamber die casting machine	143
10.3	Die casting operation: sprues, gates and runners	144

Figure no.

10.4	Die casting installation to unload, quench and dispose of part	146
10.5	Circuitry for unload, quench and disposal of part	147
10.6	Equipment layout for die cast unload, quench and trim	149
10.7	Circuitry for unload, quench and trim	150
10.8	Robot engaged in die care procedures	152
10.9	Die casting capability extended by cast-in inserts	153
10.10	Equipment layout and hand design for insert positioning	154
11.1	Typical spot welding gun used in auto body manufacture	162
11.2	Spot welding: simple minor zone change	166
11.3	Spot welding: complex minor zone change with rotation of gun and change in angle	166
11.4	Spot welding: complex major zone change on a typical auto-body weld job	167
11.5	Typical robot grouping on spot welding line	167
12.1	Typical robot arc welder	173
12.2	Electric arc welding: typical weld sequence	175
12.3	Section of tape used to break down a welding contour into small equal steps	176
13.1	Example of multiple mold produced from pattern tree	183
13.2	Special spin-control hand for 'twirling' investment casting molds	185
14.1	Plant layout for robotized chain link manufacture	194
14.2	Special gripper for chain link manufacture	194
14.3	Special hand design for holding hot metal billets	196
14.4	Robot handling cylinders in forging operation	196
15.1	Robots in Ford Motor Company press shop at Dearborn, Michigan	200
15.2	Robot in press shop fitted with two hands	200
16.1	The Trallfa spray-painting robot	211
16.2	Layout for robotized spray-painting process in automobile industry	213
17.1	The injection molding process	218
17.2	The blow-molding process	219
17.3	Plant layout for injection molding application	221
17.4	Special hand for injection molding application	222
17.5	Robot servicing two injection molding machines	223
18.1	Robot trimming steel castings as Kohlswa Steelworks, Sweden. By courtesy of ASEA	229
19.1	Typical layout for applying robots to machining applications	236
19.2	Robot tending three machine tools	237

Figure no.

19.3	Layout of three-robot line in machine shop at Xerox Corporation	238
19.4	Double-handed robot loading a lathe	239
19.5	Layout of three robots on machine line at Massey Ferguson	240
19.6	Programmable controller used with a triple robot installation at Massey Ferguson	241
19.7	Integrated robot-N.C. system for small batch manufacture	242
19.8	Double hand used in small batch machining system	242
20.1	Plant layout for robotized heat treatment line	252
21.1	Robot deburring operation at Kohlswa Steelworks, Sweden. By courtesy of ASEA	256
21.2	Further example of robot deburring. By courtesy of ASEA	256
22.1	Palletization by robot in plastics molding operation	258
22.2	Method of loading pallets to achieve maximum palletization	259
23.1	Plant layout for robotized pallet handling in brick manufacture	268
23.2	Pusher mechanism for robot arms for placing pallets	268
24.1	Layout for robot in window edge-grinding operation	272
24.2	Special double hand used in glass handling	273
24.3	Robot lifts load of glass tubes	274

Color plates (between pages 108 and 109)

1. Combined robotic and visual inspection system from Auto-Place, Inc.
2. Standard Auto-Place Series 50 robot on a double slide
3. Electrolux MHU-Senior robot engaged in heat treatment
4. Electrolux MHU-Senior robot serving injection-molding machine
5. The Cincinnati Milacron computer-controlled T^3 industrial robot in an aircraft manufacturing application
6. Two Cincinnati Milacron T^3 robots work together handling refrigerator liners
7. ASEA robot cutting ingots at Kohlswa Steelworks, Sweden
8. ASEA robots spot welding at Saab-Scania, Sweden
9. Unimate handling hot metal billet in foundry operation
10. Unimate engaged in die casting
11. Unimates in action: auto spot welding
12. Continuous path welding by Unimate
13. Stamping operation by Unimate
14. Unimate handling glass
15. Unimate line making turbine blades

Foreword

THE REAL THING

by Isaac Asimov

Back in 1939, when I was still a teenager, I began to write (and publish) a series of stories about robots which, for the first time in science fiction, were pictured as having been deliberately engineered to do their job safely. They were not intended to be creaky Gothic menaces, nor outlets for mawkish sentiment. They were simply well-designed machines.

Beginning in 1942, I crystallized this notion in what I called 'The Three Laws of Robotics' and, in 1950, nine of my robot stories were collected into a book, *I, Robot.*

I did not at that time seriously believe that I would live to see robots in action and robotics becoming a booming industry. . . . Yet here we are, better yet, I am alive to see it.

But then, why shouldn't they be with us? Robots fulfil an important role in industry. They do simple and repetitive jobs more steadily, more reliably, and more uncomplainingly than a human being could — or should.

Does a robot displace a human being? Certainly, but he does so at a job that, simply because a robot *can* do it, is beneath the dignity of a human being; a job that is no more than mindless drudgery. Better and more human jobs can be found for human beings — and should.

Of course, the robots that now exist and that are described in fascinating detail in this book that you are holding, are not yet as complex, versatile and intelligent as the imaginary robots of *I, Robot,* but give the engineers time!

There will be steady advances in robotics, and, as in my teenage imagination, robots will shoulder more and more of the drudgery of the world's work, so that human beings can have more and more time to take care of its creative and joyous aspects.

Author's preface

When Pygmalion fell in love with his beautiful creation Galatea, Venus compassionately breathed life into the marble statue and Pygmalion was blessed with an exquisite robot wife. One may presume that after the honeymoon, he put her to work. That is what this book is about, putting robots to work.

Others have and will continue to write about robot design. It is a volatile field that draws upon many technical disciplines for ever greater sophistication. To date, robots have largely been insensate, but roboticists are striving to correct this deficiency. When robots do boast of sight and touch, the list of applications (Part II), will merit a large supplement; but, meanwhile, there is much good work for senseless robots to do. The process of selecting suitable jobs and then optimizing the work place for successful economic employment of robots has been evolving since the first Unimate robot was installed to tend a die casting machine in 1961.

This author has been privy to the bulk of the successful robot installations (and to the dismal failures as well), inasmuch as Unimation Inc., with over 3000 Unimates in the field, has been the dominant manufacturer. Moreover, in application areas where Unimation Inc. experience is limited (i.e., spray painting), other robot manufacturers have been generous with application data.

Some debts should be acknowledged before attempting an exhaustive discourse on the business of putting robots to work. First of all, there was Isaac Asimov who conveniently began his prolific writing career at a tender age with robotics as a theme (thus coining the name of the science and catching the fancy of this 1940's Columbia University physics major). Then, one George C. Devol propitiously turned up at a cocktail party in 1956 with a tall tale of a patent application labeled *Programmed Article Transfer*. It was issued in 1961 as U.S. Patent 2,988,237, and good friend George went on to amass numerous other patents in robotics to the ultimate benefit of Unimation Inc.

Innovations don't happen without financial support. An imaginative entrepreneur, Norman I. Schafler, founder and still chief executive at Condec Corporation, dug down first and he was later joined by Champ Carry, then Chairman of Pullman Incorporated.

After the first industrial robot installation of 1961, there was a lot of 'hanging-in-there' to do. Not only were there remaining technical problems, but there were some formidable institutional barriers. To many manufacturing executives robotics remained science fiction fantasy. Unimation Inc. did not show a profit until 1975.

In the 60's, there were some tentative and, for the most part, abortive attempts at developing competitive robots, but none blossomed to help carry the early institutional load. Yet robotics was an idea whose time had come. By the early 1970's, the artificial intelligence community swung some of academe's attention to robotics. That interest earned support from various national governments (and today this is mounting). The Japanese jumped in with great enthusiasm; and the Japan Industrial Robot Association (JIRA) was started in 1971. Kawasaki Heavy Industries had taken a license from Unimation Inc. well before, in 1968.

In 1973, Warnecke and Schrafft of Stuttgart University wrote their book, *Industrie Roboter,* in which they uncritically catalogued every robot developed that they could unearth. They listed 71 firms as being developers of robots. By 1978, there had been some 200 efforts, most of which were abandoned. Survivors who may be taken seriously are listed in the Appendix. Only those who are in production and who back up their product with complete customer service are included in the list.

The USA could not boast of enough committed manufacturers to form an association until 1975, when the Robot Institute of America was formed. By 1978, association membership comprised 10 robot manufacturers, three robot accessory manufacturers, 25 users and three research organizations. The British Robot Association (BRA), is even younger, getting started in 1977 with strong support from academia. Little robot manufacturing has as yet been started in the UK, but research interest is strong and BRA starts out with an avant garde coterie of users and would-be users. Robot organizations are springing up throughout Europe. One of the hottest technical conference topics is robotics. The International Symposium on Industrial Robots (ISIR), is an annual event, with its venue chosen from European countries, Japan and the USA.

As with any new field, development directions are legion. Some have sought to develop sophisticated computer-controlled machines, but many more have elected to make simpler devices, with mechanical stops and pegboard programming. Robots come in all shapes and sizes, some handling only a few grams while others can cope with as much as 1000 kg. Arm coordinates can be polar, cylindrical, cartesian or revolute. Muscle power is hydraulic, pneumatic or electric.

This book considers the place of robots in factories and it considers the types available and it makes much of economics being the driving influence. An attempt is also made to predict both technological direction and the sociological implications for the last two decades of this century.

There is no question but that robotics has become an international industry, complete with all the trappings of product choice, industry association, government encouragement, public interest, a research coterie and the promise of explosive growth. Such growth of the industrial robotics industry depends upon broad acceptance of this new technology by hard goods manufacturers, a notoriously sceptical and conservative clientele. It is hoped that this book will serve to allay ill-founded concern and to eliminate some of the pain that inevitably accompanies the adoption of unfamiliar concepts.

Joseph F. Engelberger
September 1980

PART 1
Fundamentals
and management

Chapter 1

Robot use in manufacturing

Robots entered the English vocabulary with the translation of Karel Capek's play *R.U.R.* (Rossum's Universal Robots) in 1923. Capek was a Czech, and in his native language the word robot simply meant a worker. In the play, robots were the humanoid creations of Rossum and his son, constructed in the fond hope that they would perform obediently in the service of man. Now, thanks to Capek and a generation of science fiction writers, everyone knows what a robot is. The popular conception is a mechanical man, crammed full of near-miraculous components, and capable of clumsy imitations of human actions and speech. They are generally thought to combine superhuman strength with subhuman intelligence. Robots are often endowed with sinister intentions, so that the specter of a robot army marching against mankind has been a popular recurring theme in science fiction.

Fortunately, Isaac Asimov in the 1940's took it upon himself to envision robots in a happier light. Asimov's robots were benevolent. In a series of robotic stories that are both ingenious and delightful, Asimov postulated roboticists with the wisdom to design robots that contained inviolable control circuitry to insure their always 'keeping their place'. The Three Laws of Robotics remain worthy design standards:

1 A robot must not harm a human being, nor through inaction allow one to come to harm.
2 A robot must always obey human beings, unless that is in conflict with the first law.
3 A robot must protect itself from harm, unless that is in conflict with the first or second laws.

Capek gave us 'robot'. Asimov coined the name of the trade, 'robotics', and he provided all of us roboticists with an ethic.

Evolution of industrial robots

Although functioning, exact replicas of men remain technologically impossible. It is nevertheless attractive to imagine an industrial world where robots could perform all the menial, tedious, repetitive, dangerous and otherwise unpleasant jobs. Why not factories with robot labor forces? This dream becomes less far fetched as soon as we stop thinking of robots in purely anthropomorphic terms. Capek caused Rossum to say, 'A man is something that feels happy, plays the piano, likes going for a walk, and, in fact, wants to do a whole lot of things that are really unnecessary . . . But a working machine must not play the piano, must not feel happy, must not do a whole lot of other things. Everything that doesn't contribute directly to the progress of work should be eliminated.'

By considering what a man has to do at a typical machine station, it should be possible to devise a list of corresponding characteristics that a robot must possess if it is to replace him. And now a most important fact emerges. In order to increase the ratio of output to labor cost, most manufacturers have broken down their processes into small elements. Each operator has to learn only one sequence of operations, which he is then required to perform over and over again. The degree of skill is low, and there is little to learn. Thus the manufacturer is able to employ unskilled labor. And if the job has been simplified for the man, it has also been simplified for his possible robot successor — simplified, in fact, to the point where robots have become a present-day practicability.

Picture the typical unskilled or semi-skilled operative at his machine station. For the purposes of this discussion the type of machine or process hardly matters. It can be assumed that, in most cases, the operator's first job is to load the machine. Loading involves selecting the workpiece or raw materials, picking up, and placing the part or materials into the machine. Usually, it is necessary to ensure that the workpiece is correctly oriented, although there are exceptions such as objects which are spherical, or where powders or liquids are being loaded for processing. Next, a series of levers, handles, buttons or other controls have to be operated in a sequence that causes the machine to carry out its work. Then, the man unloads the machine by removing the workpiece and stacking it into a bin, or on to a shelf or conveyor. The man may be expected to perform a simple visual or mechanical check at the end of this cycle, in case the product is defective.

This simple production cycle must be repeated over and over again, to cease when the batch has been finished, the materials have run out, the machine breaks down or needs resetting, or the man goes home at the end of his shift, stops to eat, goes away for some other natural purpose, goes on strike, or simply feels tired. Sometimes, the man will continue to operate the process in spite of illness or fatigue, in which case there is a possibility of a stream of defective parts being produced before these are noticed by an independent inspector or quality controller. This situation contains the factors from which a theoretical robot specification can be compiled.

Consider, step by step, the narrative contained in the previous two paragraphs. By analyzing the human capabilities and failings involved, it should be possible to view each ingredient in robot terms and build up a list of attributes that no desirable robot should be without. This comparative exercise is best achieved by using a table, and this is shown at Figure 1.1. The left hand side of this table lists activities and other factors which are relevant to any operative working at a typical machine station. In the center column, each entry is a comment on the performance of a human operative. Possibilities for a robot substitute are given in the right hand column.

A specification for an industrial robot must obviously exclude any requirement that cannot be achieved in practice. Every feature specified must be within the scope of contemporary engineering knowledge and practice. The resulting robot must not be too big, too heavy or too expensive. It has to be reliable, and capable of doing its job for hour after hour without undue fuss. Thirty years ago these conditions could not have been met. In more recent times, the development of microelectronics, advances in computer technology, and the availability of reliable electromechanical and hydromechanical servo mechanisms have all contributed to the elevation of robots from fiction to fact.

The notes in the right hand column of Figure 1.1 provide the source material for a summary of the characteristics essential to any robot that is intended to replace a man or woman at a machine station. Here is that basic summary:

O A hand which can grip or release the workpiece
O An arm which can move the hand in three planes
O A wrist for the arm, with three articulations
O Sufficient limb power to lift and maneuver the workpiece
O Manual controls with which a person can operate the limb
O A memory, which can record manual operations
O Automatic means for controlling the limb from the memory
O Ability to function at a speed not less than that of a person
O Reliability

Physical human properties that are not essential to production, or are impracticable to achieve at the current state of the art, are not included in the basic list. Thus, the industrial robot is not given legs, because it does not need to walk in order to operate a machine. Why provide two arms if the robot can manage with only one? And, because of current technological limitations, the robot is not expected to enjoy the human senses of taste, smell, hearing, touch or sight. The last two senses will be added, and soon at that. A senseless robot is a far cry from the robot of popular fiction, but it is the design concept on which several kinds of robots have been built and deployed profitably in the manufacturing industry.

It is a mistake to limit the robot specification to a contrivance that

FUNCTION	HUMAN OPERATIVE		ROBOT OPERATIVE	
Select workpiece	H1	Uses senses of sight and touch.	R1	Visual methods are prohibitively difficult at the present state of the art. The practicable methods are restricted to sensing physical contact, and by a pre-programmed command which directs the robot to the workpiece location.
Picking up	H2	Uses combination of arms, hands and body. May need mechanical assistance with heavy lifts.	R2	Strictly, the robot only needs one arm and a hand, although these must be jointed to allow the hand to move in three dimensions and to swivel. The use of hydraulic power can produce greater lifting capacity than a man's.
Placing	H3	Similar to H2, but uses his eyes to ensure that the workpiece is correctly orientated.	R3	Similar to R2, but without the sense of sight, orientation becomes difficult. A robot operator needs workpieces that are presented to it in some pre-defined and constant attitude.
Operate machine	H4	Uses any one or all of his five senses to follow the operation of the machine and activate the controls as necessary. Has a memory, with which he can learn the sequence and timing of operations.	R4	Because it is not able to see, hear, or otherwise witness the progress of the machine, a robot must be pre-programmed to carry out its operations according to a timed sequence. A man has to do the teaching, and the robot has to have an internal memory to store the information. Computer technology has made this possible.
Unload machine	H5	This is similar to H2.	R5	Similar to R2.
Stack the workpiece	H6	This is similar to H3.	R6	Similar to R3, but orientation is less of a problem, because the machine will usually be one which holds every workpiece in the same position and attitude.
Inspect the workpiece	H7	A man should be capable of inspecting, as required, using his senses of sight and touch, or by using gauges and other measuring equipment. He is able to speak, and to tell the foreman of any machine problem.	R7	A robot would be capable of inspection by automatic gauging, by probing, and by telemetry or datalogging. Although not able to speak, the robot could be designed to give an audible or other warning of any defect in the workpiece.
Machine breakdown	H8	Able to use any one of his senses to perceive possible trouble. For example, he could recognise the smell from a hot motor, and switch off the machine.	R8	A robot would be capable of detecting a machine breakdown only where this produce defects in the workpieces that the robot could detect, or where telemetry points were set up in the machine.
Machine setting	H9	This operative has been defined in the text of Chapter 1 as unskilled or semi-skilled. He must, therefore, expect to have to call for skilled assistance whenever the machine needs resetting or retooling.	R9	The robot could be programmed to stop after a period calculated to allow for tool wear. It is also practicable to command the robot to probe the workpiece and/or the machine for broken drills, taps, reamers and other tools. A skilled human operator is needed to carry out rectification and resetting.
Shift periods	H10	A man has to leave for home after a shift period of between eight and twelve hours. He has to be mobile – able to walk and run.	R10	There is no need for a robot to be self-mobile. It can be fixed to the floor, and commanded to work round the clock every day. Where three shifts are operated within 24 hours on one machine, a robot operative can take the place of three men — more men if the robot can be set up between two or more adjacent machines that it can operate in sequence.
Workbreaks	H11	A man has to take time off work for eating, and for other physical needs.	R11	Robots do not need time off for eating or for other natural functions. They will be subject to down time for maintenance and, however reliable the design, there will inevitably be occasional breakdowns.
Strikes, go-slows and overtime bans	H12	Collective disruptive action from the labor force is a problem that can hit any production plant.	R12	A robot will obey commands slavishly and there is no labor organization for robots.
Illness and fatigue	H13	Men get tired, sometimes feel below par, and they may be absent from the workplace when they fall sick. These conditions can be aggravated or caused by unpleasant working accommodation (heat, noise, toxic fumes, smells, cramped space, etc).	R13	Robots can be expected to break down very occasionally (see R11). But it is possible to design robots that can operate in extreme conditions which might damage a man's mental or physical health — such as noise, heat, vibration, noxious fumes, smells, cramped work space and even radioactivity.

Figure 1.1 *Comparison of human and robot characteristics*

A robot is not capable of performing all the tasks that its human counterpart can achieve. Conversely, robots are able to do some jobs better than men, especially where these demand repetitive work for long periods under arduous conditions.

can simply attempt to imitate its human prototype. We know that, within the limitations of current technology, any real life industrial robot is going to be a machine that lacks many human senses and sensibilities. As mere human imitators, therefore, robots are likely to be very inferior models. But robots can be conceived with other capabilities, which are not anthropomorphic, and which can compensate in many respects for their inability to see and feel.

Robots should be capable of outperforming men in hostile working environments, where noise, vibration, smells or danger act adversely on the physiological system. Robots do not have to stop for eating. They have no wives to go home to. They do not get tired, they can work right round the clock. Unlike men, robots are not going to be gregarious creations, and we should not expect to find them participating in drama groups, stamp collecting societies, sports clubs or the like.

The use of hydraulic or electric power gives a robot more potential muscle than a man or woman. The design specification should take all these factors into account, so that the resulting product is extended into a robot which exploits its mechanical advantages as fully as possible.

Robot controls became feasible through the advent of digital logic and miniaturized solid state electronics — the very same techniques used in computers and in numerically controlled (NC) machine tools. This common root means that communication links can be established between robots and computers, and between robots and NC machines. This opens up exciting future possibilities for operating robots directly from computers, or for synchronizing them with NC machines. Companies which use computers for design (CAD) might even be able to have their computer design the product and produce the robot control program, so that not only the human machine operator is replaced, but the draftsman as well. These prospects are discussed in later chapters, but they illustrate the need to compile a specification which, while avoiding unnecessary human attributes, contains all the advantages that robots can offer.

Such a specification can now be drawn up. By starting from human performance, by taking Capek's advice to strip away factors not essential to the progress of work, and then by adding facilities available from today's technology, we arrive at the extended specification shown in Figure 1.2.

Near relations of the robot

Even though everyone knows what a robot is, this presumptive knowledge is richly diverse. Therefore, a book devoted to the use of robots ought to attempt definition, and then put to bed related technologies that don't qualify and that therefore will not be considered herein.

1	A hand, capable of gripping and releasing the workpiece
2	An arm, which can move the hand in three planes
3	A wrist for the arm, with articulations that allow the hand/wrist assembly to be aimed anywhere in the workspace
4	Sufficient muscle power to lift a 500 pound (225 kilo) workpiece
5	Positioning repeatability to 0.3 mm
6	Manual controls, with which a person can operate all robot limb functions
7	A built-in memory which can learn the human teacher's instructions *Ulrich*
8	Automatic systems which enable the memory to control operations in the absence of the human teacher
9	A speed of operation which is at least as fast as a person *Rehg. P. 19*
10	A library of programs which can be selected at will, allowing the robot to be switched back to operations that it has been taught in the past
11	Facilities for safety and process interlocks with the plant or machinery that is to be operated
12	A computer compatible interface
13	Reliability of at least 400 hours MTBF (Mean Time Between Failure) in the actual working environment
14	Configuration which allows easy maintenance, with quick access and interchangeability of parts in the event of breakdown, aided by self-diagnostic routines

Figure 1.2 *Robot characteristics — extended specification*

Webster's Seventh New Collegiate Dictionary defines the robot as 'an automatic apparatus or device that performs functions ordinarily ascribed to human beings or operates with what appears to be almost human intelligence.'

The Robot Institute of America attempts to be more precise: 'a robot is a reprogrammable, multifunctional manipulator designed to move material, parts, tools or specialized devices, through variable programmed motions for the performance of a variety of tasks.'

One feature which a device must possess if it is to rank as a robot is the ability to operate automatically, on its own. This means that there must be inbuilt intelligence, or a programmable memory, or simply an arrangement of adjustable mechanisms that command manipulation. For these reasons, the following devices cannot be classed as robots, or part of the subject of robotics. They are listed here because, in many cases, they do share some of the technology that does apply to robots, and they are therefore related.

Prostheses: artificial replacements for parts of the human body. Limbs are of special interest, because they use servomechanisms and linkages to achieve movement and control to replace that of the missing or useless member, and some employ advanced electronics to amplify and harness body electrical impulses that were originally intended by Nature to stimulate human muscles.

Exoskeletons: frames which surround human limbs, or even the whole human frame, and which amplify the available power of the man or woman. They do not have intelligence or memory of their own, and so are not robots. In fact, it is most important that they cannot operate independently, since very accurate control and response times are absolutely essential to the safety of the user, who could easily damage himself or his surroundings in a badly designed system.

Telecherics: the name given to the subject of remote manipulators. These are used to add distance to the motions of a human limb, so that the operator can work outside the environment in which work has to be done. One obvious example is the range of devices used to handle radioactive materials. Once again, some of the linkage systems used may be similar to those used in some robots, but until the human operator has been replaced by some artificial intelligence, operating within a closed servo loop, manipulators are not robots.

Locomotive devices: imitate men or animals by walking on legs instead of the more usual mechanical method of wheels. The technology is very closely related to robotics, and to exoskeletal systems. A mobile robot is, of course, a locomotive device plus some form of built-in intelligence. Although experimental vehicles have been made, with up to eight legs, their intended advantages over wheeled or tracked vehicles for traversing rough, uneven ground have yet to be commercially proved.

These four classes of near relations are illustrated in Figure 1.3.

Robot cost *versus* human labor

At the beginning of this chapter, it was said that one of the most important factors which has paved the way for robots has been the changes in manufacturing methods to allow employment of unskilled people. In the interest of production efficiency, work has been broken down into a series of simple repetitive tasks that can be taught quickly. In fact, much factory work has been reduced to activities that are grossly subhuman, and there are few artisans to be found in modern manufacturing plants. This applies not only to the assembly line labor that Charlie Chaplin championed in his classic *Modern Times,* but also to fully automated machine stations, where the only human activities left are loading and unloading.

Charlie Chaplin's message was that unskilled and semiskilled workers were being exploited in subhuman work so that the United States' appetite for hard goods could be satisfied. In 1936, when *Modern Times* made its protest, labor was plentiful, cheap and more easily intimidated. The humane treatment and use of these human beings was

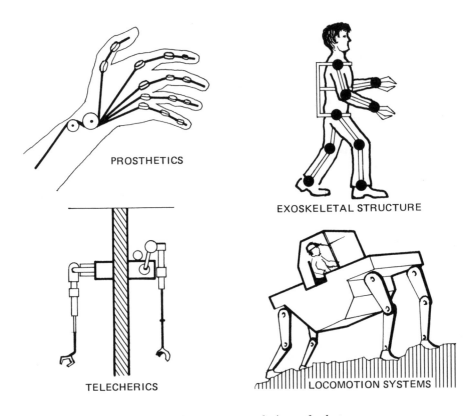

PROSTHETICS

EXOSKELETAL STRUCTURE

TELECHERICS

LOCOMOTION SYSTEMS

Figure 1.3 *Some near relations of robots*
The devices illustrated here have some technology in common with industrial robots. They cannot themselves be classed as robots, however, because none of these examples contains any inbuilt intelligence. These are certainly very sophisticated tools, but they are not able to work independently without constant directions from a human being.

a noble goal; but the captains of industry were motivated less strongly by noble goals than by the demand for profits. There was simply no economic justification or need to replace obedient and cost-effective human operatives with robots.

Many things have changed. A quarter of a century later, the demand for semi-skilled labor had increased. A semi-skilled factory laborer in the U.S.A. was costing $6,000 a year in pay and fringe benefits. In the automotive industry, this cost was destined to rise to $30,000 by 1979, and the trend continues: see Figure 1.4. The technique of designing work to eliminate skill had backfired. Workers demonstrated their aversion to dull, repetitive jobs by demanding high pay; higher even than that earned by workers in more skilled and more interesting jobs. Furthermore, turnover and absenteeism tended to follow job dissatisfaction, adding to the total cost of so-called cheap labor.

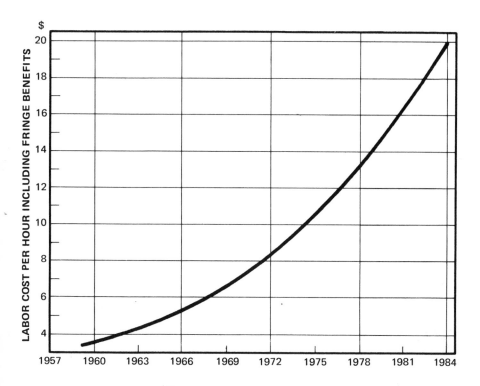

Figure 1.4 *Labor cost escalation in the U.S. automotive industry*
The trend of increased payments to workers continues, so that labor which was once regarded as cheap now has to be paid for at rates which compensate for the monotony and unpleasant nature of the job.

Economic and social factors governing the employment of labor are very much interdependent. These factors are discussed in later chapters when the advantages to be gained from introducing robots are evaluated. Several methods exist for analyzing the results of capital expenditure on new machinery and plant but, for the purposes of this chapter, one simple approach (the payback method) is sufficient to demonstrate that the economic viability of industrial robots is a fact. Not only are robots possible to make — they have become an economic alternative to human labor.

Payback analysis

Suppose that a human operator must be paid $24,000 a year in wages and other benefits to do a particular job that could be done, instead, by a robot. Suppose also (for simplicity in this case) that the robot would produce the same rate of output as the human, per shift. The useful working life of this robot is going to be at least eight years, but a pay-back period of under three years is normally sought by accountants. The robot and its accessories are going to cost $55,000, and the annual maintenance costs have been demonstrated to be $3000. One other important factor is that the robot would be capable of working more than one shift, so that it could in fact replace at least two men.

Payback formula

$$P = \frac{I}{L - E}$$

where

P = the payback period, in years
I = the total capital investment required in robot and accessories
L = total annual labor costs replaced by the robot
E = maintenance costs for the robot

Calculation for one shift working

I = $55,000
L = $24,000
E = $ 3,000

Period $= \dfrac{55,000}{24,000 - 3,000}$ or 2.62 years

For two shift working

L becomes $48,000 and
E rises to $5,000

The payback period reduces to

$\dfrac{55,000}{48,000 - 5,000}$ or 1.28 years

Figure 1.5 *Payback evaluation of robot costs*
Application of the simple payback formula demonstrates that robots can be viable in replacing expensive human labor. The cost benefits are seen to be even greater when two shift working is in operation, in this example reducing the payback period from 2.62 to only 1.28 years.

The calculations are displayed in Figure 1.5. For one shift operation, the payback period works out at 2.62 years. If the robot is utilized more fully, over two daily shifts, the payback period drops to only 1.28 years. This result is likely to satisfy even the most critical accountant. The figures become even more impressive when it is realized that the effective working life of industrial robots can be more than eight years, based on actual operating experience, and that they can work three shifts right round the clock.

Die casting — an early success story for industrial robots

The following example dramatizes the potential in robot application. The process of die casting is carried out to produce accurate, well finished workpieces by the high pressure injection of molten metal into an accurately machined steel die. The most typical metals used are zinc, aluminium, and their associated alloys. The process is described in detail in chapter 10, but here is a brief summary of the process.

The die consists of two separable, mating halves. Before use, all residue from the previous casting operation has to be cleaned off, and then both halves must be lubricated. A press brings the die halves together, and the charge is injected under high pressure. Pressure is maintained for a period sufficient to ensure that the die has been completely filled. The pressure of the charge results in a slight escape of metal through the mating die surfaces, which has the appearance of thin fins projecting from the workpiece when it is first removed from the die. These protrusions are called 'flash'.

It is necessary to water-cool the die and its contents to solidify the molten metal. Once the metal has solidified, the press opens the die halves and the casting can be removed. The die casting usually needs further cooling in a quench tank, and then the flash must be removed. This is done in a trim press.

Automating the die casting process

Die casting can be an unpleasant occupation for human operatives, combining tedium, heat, noxious fumes, and other undesirable problems in the working conditions with actual physical danger. In common with many other industrial processes, output rate, product quality and the amount of scrap produced all depend to a large extent on the skill of the individual operator. Die casting has, perhaps, more than its fair share of operational variables, ranging from the chemical and physical state of the hot metal, through mold treatment, machine adjustment, cycle timing and the physical and mental condition of the human operative at any given time.

Most die casting shops have to handle small batch quantities of workpieces. Generally speaking, any attempt at special purpose automation would prove to be prohibitively expensive in such circumstances. To overcome many of the variables associated with human operation, some sort of automation is desirable. But this has to be flexible. It must be possible to set up new process programs easily and cheaply as each new batch is loaded. Better still, when a repeat batch is needed in the future, it should be possible to recall the appropriate program from a memory, rather than have to start programming all over again. This is the sort of flexible automation that industrial robots can provide, and this is one of the reasons why die casting was an early commercial success story for robotics. It was, in fact, the very first task taken on by a robot, as far back as 1961.

Robotized die casting

Robots can be used at any stage of the die casting process, or in trim press operation. Some applications are suitable for complete systemization, with one or more robots handling all operations from the start of the die casting machine cycle to stacking trimmed castings. In this instance a straightforward one press-one robot set-up serves to illustrate the flexible automation that robots can provide.

Picture the robotized press. In place of the human operator stands the robot, with its single, powerful arm. This arm can be extended or retracted, raised or lowered, and it can be swivelled from side to side. At the end of the arm is a mechanical wrist which can also be swivelled. Attached to the wrist is a hand capable of grasping the workpiece. A control panel on the robot body is supplemented by a pendant control box, from which a human operator can control the movements of robot arm, wrist and hand. The pendant fits neatly into the man's hand. All the pendant controls are designed logically, to produce robot movements that correspond with the directions in which the control knobs are themselves moved.

The robot is able to record each instruction from its operator in a stored program. The control unit has provision for interlocks between the robot and the machine which it is going to operate. These interlocks are essential and ensure that no part of the production cycle can be initiated before the previous process stage has been satisfactorily completed. This system of feedback from machine to robot depends on limit switches and other sensors that can pick up the travel or positions of moving parts. For example, the travel of the injection piston can be used to provide indication that the mold has been completely filled. An infra-red detector senses the presence of a hot die casting in the robot hand; absence of sufficient heat radiation would signal that all or part of the workpiece had been left in the die.

In many respects the electrical links, interlocks, limit switches, temperature sensors and their associated circuitry resemble those used in special purpose automation controls. There are, however, essential differences. The set-up or teaching time is far simpler and quicker using a robot. And all these process controls are fitted to standard, general purpose machinery. When one batch of die castings has been finished, another pair of different dies can be used in the same die casting machine, using all the same equipment and controls. True, the robot program has to be changed, but this is either accomplished by a simple reteaching operation, or by re-using a previous program that has been stored on a tape cassette. Had this all been a special purpose, custom built machine, the whole investment would have to be written off when the product run finished.

Summarizing, we have a general purpose die casting machine, operated by a single armed robot equipped with teaching controls and a memory. There are logic links between the robot and the machine, and a human operator can be called upon to carry out initial robot instruction. The usual noise, heat and fumes are there too, but the man knows he can get right out of there just as soon as he has taught the robot to make die castings. The robot memory ensures that the robot will carry out every operation for hour after hour, shift after shift, without fatigue and without variation. Robot users have reported significant reductions in scrap and reject rates. Machine utilization rates are increased — one typical user reporting 25 per cent. Removing the man from the immediate operational vicinity means reduced needs for safety equipment, with consequent savings in costs. And there are obviously dramatic reductions in labor costs, because there is no longer any need for an operator to be kept fully occupied at the machine. All of these advantages add up to a powerful argument for robotization.

Robots *versus* special-purpose automation

It is not enough to compare the robot with its human counterpart. Long before our technological resources permitted the development of an industrial robot, we had produced automatic machinery that was capable of mass-producing many of the products which are in broad commercial and consumer use. For example, there are machines that make bottles and other machines that fill and cap these bottles. There are machines that automatically manufacture our light bulbs. And, there are machines called transfer machines that can ingest a raw casting and deliver a completely machined engine block. All of this equipment has come to be lumped together as automation.

It is reasonable to conclude that robotics is no more than a subclass of automation and, certainly, today the production of robots is an almost negligible percentage of industry's total expenditure for automation equipment. However, there is something about the robot mystique that singles out robotics for special attention. As robots become ever more sophisticated, the distinction will become more dramatic.

A robot that can boast of eyesight and tactile sensing will certainly be better at emulating a human operator. And, the robot carrying out its tasks in anthropomorphic fashion reminiscent of a human worker is certainly distinct from an aggregation of mechanisms specially designed to produce one or another specific product.

We have noted that the prime advantage a robot has over human labor is economic. But the robot also has advantages over special-purpose automation. In fact, the robot increases its niche in industry by driving a wedge in between manual labor and special-purpose or 'hard' automation.

The robot can take over from human operators because its flexibility permits it to accomplish tasks that are not readily accomplished by special-purpose automation. It is axiomatic, of course, that these activities are done more economically by robots than they would have been done by humans. But how is it that a robot can also intrude on activities that might be suitable for hard automation? The answer is that an industrial robot provides three unique benefits when compared to hard automation.

1 Reaction time: a sufficiently flexible and sophisticated industrial robot is 'off-the-shelf' automation. Such robots, by definition, are manufactured for stock and not to special purpose order. They are poised and ready to go.

Evidently, when a decision is made to automate, the financial benefit does not start until the automation is operating. Thus, fast reaction time provides a cost saving which must be credited to the industrial robot when comparing investment with the cost of special purpose automation.

Consider the example of the automobile manufacturer who, through the use of industrial robots, was able to design the automated assembly line concurrently with the design of the vehicle itself.

The ability to shorten the cycle time for introduction of new models is most highly valued by the automobile industry.

There is another type of reaction time benefit to be derived from an easily taught industrial robot. One U.S. job shop diecaster with 16 Unimates tells how important rapid programming is when short runs are the rule. And, how convenient it is to touch up programming virtually on the fly to compensate for process variability such as die deterioration.

2 Debugging: when literally millions of dollars are spent in design of a sophisticated industrial robot slated for replication by the hundreds (perhaps by the thousands), the debugging investment is on a comparable scale. Such fastidious testing and field trial is economically impractical for special purpose automation. The experience of General Motors in Lordstown bears this out. Twenty-six Unimates went to work there amidst grave misgivings regarding the electronic-hydraulic sophistication of this equipment. But the real debugging anguish came in getting the conventional automation on stream — all those simple indexing devices, solenoids, actuators and clamping mechanisms that are the legacy of special purpose assembly systems; those single purpose elements whose debugging in a system occurs on the production floor, after delivery.

The more flexible the industrial robot, the smaller the percentage of a system that must be purpose built and the lower the debugging cost becomes. This principle applies right down to individual work stations. If manipulative power is limited, it usually is paid for by additional set up cost, equipment moving cost and lost time.

3 *Resistance to obsolescence:* the essence of robotics is to provide a form of automation which is immune to obsolescence. A proper industrial robot is neither product, nor operations, nor industry limited. The experience of Unimation Inc. proves the point. With thousands of machines in the field, some having worked for 15 years accumulating over 90,000 hours in production, there is still no sign of technological obsolescence. Unimates get overhauled, they get new job assignments, they get sold in an after-market, but they do not get scrapped. Model change has not embarrassed the Unimates in the field to date and there are no ominous signs.

Chapter 2

Robot anatomy

Robot capabilities range from very simple repetitive point-to-point motions to extremely versatile movements that can be controlled and sequenced by a computer as apart of a complete, integrated manufacturing system. It would be convenient if all robots could be placed into neat categories, each category being labelled with all the job capabilities belonging to its robot members. Although this can, and will, be attempted here, there are dangers and pitfalls in trying to establish a rigid robot classification. The real problems result from the fact that, while a simple robot might be perfectly capable of doing a good job in a plant, a more sophisticated (and therefore more expensive) robot could possibly do the job even better, and even more profitably. Classification itself is not too easy, since robot design development progress has caused a good deal of overlap between what were once clearly distinguishable robot types. Rather than be defeated by these arguments, we can first consider three simple robot categories, and then go on to look at the complications introduced by variations in anatomical conformation and different designers' preferences in matters of drive mechanisms and control techniques.

Robot classification

All robots consist of two major component systems. First, there are the moving parts, chiefly comprising the arm, wrist and hand elements. This is the most obvious part of the robot to an outside observer. The moving system is often referred to as the manipulator, but this term can be misleading, because it is easily confused with one of the robots' near relations, the telecheric device (see Chapter 1).

Complementary to the moving robot system is the control system. At its very simplest, this might consist only of a series of adjustable mechanical stops or limit switches. At the other extreme are the computer type controls, which give the robot a programmable memory, which allow the robot drives to follow a path that is accurately defined all along its length by a series of continuous coordinates, and which can also be coupled with another computer or machine control system to synchronize the robot for the most efficient and safe production operation possible.

Comparison between any two robots that belong, broadly speaking, to one of the three categories listed below could easily reveal that quite different drive systems had been employed to achieve roughly the same ends. Control systems are likely to correspond more closely between robots in the same category. Discussion of different drives and control possibilities follows later in this chapter.

Limited sequence robots

As its name implies, a limited sequence robot is at the least sophisticated end of the robot scale. Typically, these robots use a system of mechanical stops and limit switches to control the movements of arm and hand. Operation sequences can often be set up by means of adjustable plugboards, which are themselves associated with electromechanical switching. By electromechanical switching is meant a combination of relays and rotary or stepping switches. As a result of this kind of control, only the end positions of robot limbs can be specified and controlled. The arm, for example, can be taken from point A to point B, but the path in between is not defined. Thus, the controls simply switch the drives on and off at the ends of travel. This mode of operation has earned such machines the nicknames of 'pick and place' robots or, perhaps impolitely, bang-bang machines.

Figure 2.1 is a block schematic diagram of a limited sequence robot. The drive mechanism could be electrical, pneumatic or hydraulic. Most robots of this type are small, and tend to move faster than their bigger, more complicated brothers. The use of mechanical stops and limit switches gives good positional accuracy, which is typically repeatable to better than ±0.5 mm. Purchase price is around 25%-50% of that required for bigger robots. They have been used successfully in a variety of applications, including die-casting, press loading, plastics molding and as part of special-purpose automation. Disadvantages, other than the obvious control limitations, are that the number of limb articulations is likely to be few, and setting the machine up is more time consuming and tedious than for those with better control systems. The number of movements possible in a total production sequence must be limited to the number of limit switches, stops, and programmable switches contained by the robot. Such robots are not 'taught' to perform their job, but have to be set up in the same way as an automatic machine would be adjusted. There is no memory unit as such, other than that embodied in the settings of the plug board and all the mechanical stops.

Unlike robots in the other categories, the simple limited sequence control system cannot exercise any real control over the limbs while they are actually in motion. It is possible to provide more than one stopping point along each path, but the primitive nature of the memory

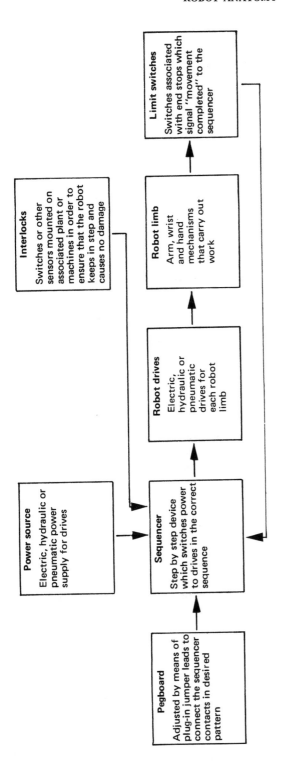

Figure 2.1 *Schematic arrangement of a typical limited sequence robot*

The program sequence is predetermined by arranging plug-in jumper leads in an appropriate pegboard pattern. Each robot limb movement is set to start and stop by means of adjustable end stops. Limit switches at each of the end stops operate to inform the sequencer that particular movements have been completed. The sequencer, controlled by signals from the limit switches, rotates step by step to supply power to each robot drive according to the requirements of the production operation. Interlock switches on the operated machine and other associated plant ensure that the robot is kept in step with the manufacturing cycle, and prevent the robot from disgracing itself by causing accident or damage by collision.

system restricts the number of these for practical purposes. In this type of machine the motion of the working arm is often extremely limited also, as are the number of available degrees of freedom. In fact, with such simple machines, options such as wrist and hand movements and arm rotations are provided to meet the needs of specific customers.

Programming limited sequence robots is usually accomplished by setting up all the end stops in their appropriate positions, and by adjusting contacts in the sequencing unit so that all the steps take place in the correct order. A peg board is often used to make the task of sequencing quicker.

Here is a description of a limited sequence device in operation. When a sequence is to be started, the controller has to switch power to the relevant drive motor. If the drives are electric, then the controller will probably close a relay to switch the current through. Where the drives are hydraulic or pneumatic, then appropriate solenoid valves are operated. The motion generated by the drive normally continues until the moving limb is physically restrained by hitting an end stop, the physical shock being cushioned by some form of shock absorbing device. Thus there are only two positions at which the moving part can come to rest, one at the beginning and the other at the end of a programmed move. Obviously, the system is arranged so that a limit switch cuts off the motive power as soon as the end stop is reached. When the initial movement has been finished, the limit switch not only cuts off the drive power, but it also signals the controller that the particular movement has been finished, so that the next movement can start.

The controller is a sequencing or stepping switch. It could be a set of contacts operated by cams on a spindle which can be rotated in steps of a few degrees at a time by a small electric motor. Each time the sequencer receives its own drive signal, it steps to its next position, and so switches drive power on or off to a particular part of the robot. This performance is repeated, step by step, until the whole program has been carried out, the manufacturing cycle is complete, and the robot is ready to start all over again on another cycle.

So far so good. But how does the controller ensure that the robot doesn't put its arm into the closing jaws of a press, or try to load a workpiece into a spinning chuck? The robot cannot see the machine it is trying to operate. There are no robot senses equivalent to those of a human operator. Some method has to be found to make the robot aware of the real world around it. This is accomplished by providing additional limit switches or other electrical sensing devices on the machine to be operated. These are connected to the controller to provide additional signals to the sequencer, complementary to those obtained from the switches mounted on the robot itself. Robot limb movements are therefore carefully interlocked with the machine being operated. This prevents the robot from trying to commit suicide, avoids collision damage to associated plant, and enables the robot to carry out its operations not only in the correct sequence, but also at the appropriate moments in time.

Playback robots — with point-to-point control

One characteristic of limited sequence robots is that they are generally difficult to reprogram. This arises from the nature of the control system and memory, which are all embodied in a complex and interdependent set of limit switches, interlocks, end stops and electrical connections. Not only does this kind of electromechanical arrangement prove tedious to change, but it also limits the number of different sequence steps that can be accommodated practicably within the control system.

Another method for achieving positional control of each limb relies on the provision of some form of servo mechanism. Figure 2.2 illustrates the principle in simple diagrammatic form. Each movable robot limb is fitted with a device that produces an electrical signal, the value of which is proportional to the limb position. The system is arranged so that the direction of drive travel is such as to reduce the positional error (obviously), and as the limb moves closer to the desired position, the error signal automatically reduces until it reaches zero, and the limb stops in the correct position. This is analog control, and in practice calls for a high degree of engineering skill in design to achieve satisfactory positional accuracy, and freedom from oscillation.

If a knob is provided on a control panel which can vary the command signal for a particular limb, then that limb will move as the knob is moved. Thus, a form of remote control is achieved, and the control panel can be given as many knobs as there are limbs to provide a man with the means for operating the robot.

The device described so far is a manipulator, and not a robot. A memory unit has to be added to complete the control unit before it can properly be called a robot and earn its place rightfully in this book. Once the memory unit exists, a very flexible robot results. The position of the limbs at each operational step, and the total operational sequence are all recorded in the memory. The memory is then used to stimulate all the servo systems. The procedure for setting up such a robot is far easier than for a limited sequence robot. It is only necessary to use the controls to drive the robot limbs to the required position for each operational step, and then to record the exact condition of the robot in the memory by the simple act of pushing a button before proceeding to drive the robot to the next step in the sequence. In other words, the robot can be taught, by simply driving it through all stages of the operation.

For obvious reasons, a robot which can be taught in this manner is sometimes known as a playback robot. It is still essential to provide

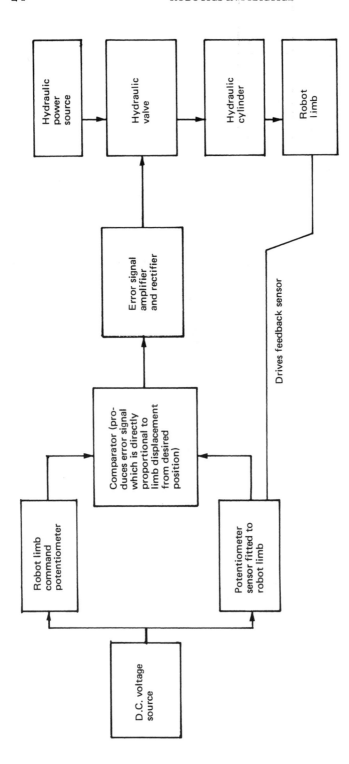

Figure 2.2 *Analog servo system*

This diagram illustrates one conventional way to achieve positional control of a robot limb.

safety and control interlocks between the robot control unit and the machinery being operated, to prevent collisions and other problems. Not only is the robot far easier to set up than a limited sequence robot, but the memory unit is able to take advantage of modern technology by digitizing all of the command position data. This means that the robot is able to remember a large number of steps.

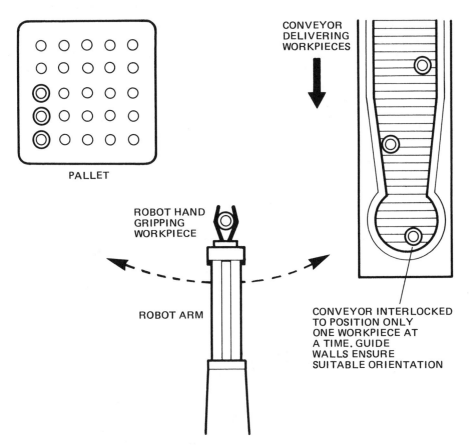

Figure 2.3 *Task for a playback robot with point-to-point control*
In this example, the robot is required to pick up cyclinders from the end of a conveyor system, and transfer them one at a time to the pallet. The conveyor is specially equipped with side wall guides and a photoelectric interlock to ensure that cylinders arrive at the pickup point singly and upright.

The problem set out in Figure 2.3 serves to illustrate the operation of a playback robot capable of point-to-point control. The robot is required to pick up cylinders from the end of a conveyor, and load them on a pallet fitted with 25 spindles. A photoelectric control on the conveyor ensures that the conveyor stops as soon as a cylinder has been delivered to the pickup point. Otherwise, there could be a log jam which might confuse the blind robot. Guide walls coax the cylinders into a well-defined area, so that the robot can always be certain of finding a cylinder in the same place, and in the same attitude (standing on one end). Location of the pallet is also important, and this is arranged on a table provided with locating lugs. When the full pallet is eventually removed and replaced with an empty one, these lugs position the new pallet accurately in the same position occupied by its predecessor. Although the robot is always going to pick up cylinders from one point, it is being asked to set them down in a sequence of 25 different locations, one for each spindle on the pallet. A problem of this size demands a robot memory far greater than that embodied in the electromechanical control unit of a limited sequence device.

Figure 2.4 is a diagrammatic representation of the robot to be used

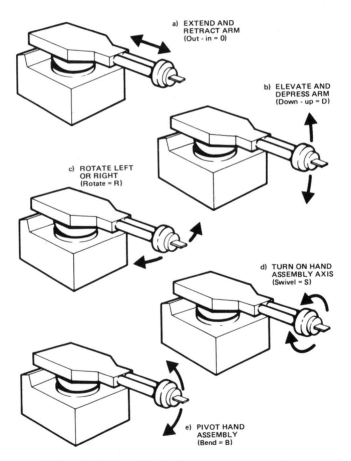

a) EXTEND AND
   RETRACT ARM
   (Out - in = 0)

b) ELEVATE AND
   DEPRESS ARM
   (Down - up = D)

c) ROTATE LEFT
   OR RIGHT
   (Rotate = R)

d) TURN ON HAND
   ASSEMBLY AXIS
   (Swivel = S)

e) PIVOT HAND
   ASSEMBLY
   (Bend = B)

Figure 2.4 *Typical articulations of a playback robot with point-to-point control*
There are five articulations in this case, all of which are programmable.

for this job. The human operator will lead the robot through all the steps of the operation, one simple step at a time, recording each move in the robot memory as he goes. To make his job safer and easier, the robot can be switched to a special 'teach' mode during this process, which reduces the robot's operation speeds, and gives full command of all robot movements to the teacher. The operator is provided with a control panel mounted on a pendant cable, which allows him to stand in the most favorable position for observing and controlling the action. One of these teach pendants is illustrated in Figure 2.5.

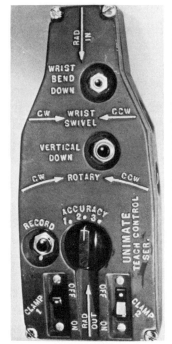
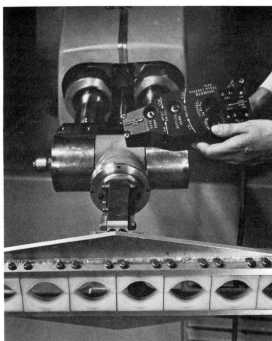

Figure 2.5 *Teach pendant for instructing playback robot with point-to-point control* This is a pendant control unit which can be held in the hand by a human operator while he takes a robot through its learning process.

There are many variations in the approach to positional control with different limb position sensors and different power sources. For any servo system chosen, dynamic performance demands are harsh. No matter what the arm attitude or its load, high-speed and critical damping are urgently required.

In order to prevent the robot from taking an inappropriate action two interlock switches are essential. One of these is arranged at the pallet, to prevent the robot from trying to load cylinders before an empty pallet has been put into place. The other interlock is taken from the same photoelectric device which controls the conveyor. Just as it is essential to stop the conveyor from depositing further cylinders at the pickup point as long as a cylinder is at that position, so it is necessary to stop the robot from attempting a pickup when no cylinder awaits it. Otherwise, we should be able to observe the spectacle of the robot grabbing at thin air, and trying to deposit an imaginary cylinder onto the pallet.

The operator must arrange for a cylinder to arrive at the pickup point, and he has to ensure that this is in the correct vertical attitude. Using the pendant controls, and with the robot switched to the teach mode, the training process now proceeds, step by step, as follows:

1 Move the robot arm until the gripper is just above the cylinder to be picked up, with the hand open.
2 Rotate the hand and wrist controls as necessary to align the grippers level, in a horizontal plane.
3 Record the robot attitude by pressing the 'record' button.
4 Lower the arm until the grippers surround the cylinder. Carry out fine adjust until the grippers are level, and symmetrical about the cylinder.
5 Press the record button.
6 Close the hand, so that the cylinder is gripped.
7 Press the record button.
8 Raise the arm well clear of the conveyor, so that it is above the height of any obstructions between the conveyor and the pallet.
9 Record.
10 Swivel the robot and adjust the arm extension until the cylinder is positioned approximately above the first peg on the pallet.
11 Record.
12 Lower the arm carefully until the hole in the cylinder is just above, but not touching, the first peg in the pallet. Ensure that the cylinder is held vertically, and if necessary, adjust the hand, wrist and arm controls to achieve this.
13 Record.
14 Lower the arm until the cylinder rests on the pallet.
15 Record.
16 Open the robot hand.
17 Record.
18 Raise the robot arm clear of all obstructions.
19 Record.
20 Repeat this operation for all remaining cylinders, until the pallet has been filled.
21 Replace the pallet with an empty one, switch the robot to normal operational mode, switch on the conveyor, push the robot start button, and watch the action.

Provided the interlocks have been correctly set, and given a regular flow of cylinders along the conveyor, the robot should now proceed to carry out the filling of the pallet in a regular, reliable and untiring manner. The robot will be seen to stop momentarily at each of the positions where the record button was pressed during the teaching session.

It is well to realize exactly how the robot motion operates. When the robot is commanded to move from one position to another, this could involve independent operation of two or more of its articulations. The only information which the robot has been taught concerns the attitude of all limbs at the start of the move, and the new attitude of those limbs when the particular move has been finished. While making the move as fast as it can, and while moving all limbs simultaneously to fulfill the given command, there is no definition of the paths which the robot limbs will trace. In other words, the control is point-to-point, as the name of this class of robots indicates. This is why it was necessary for the operator to record some of the intermediate robot positions in this example. If step 8 had not been recorded, for example, the robot arm could have clouted any obstruction between the conveyor and the pallet. Steps 10 and 12 were recorded to prevent the robot arm coming in obliquely to the pallet, which would have caused the cylinder to strike the pallet pin at an awkward angle, thus causing general mayhem. Each vital turning point in the robot trajectory must be recorded in a point-to-point control system.

Robots with point-to-point control operating on a digital basis may have virtually unlimited memory. In the example just described, the human teacher has programmed in nine separate record points for each cylinder to be handled. This amounts to 225 recorded steps for all 25 cylinders. Point-to-point robots are obviously capable of doing any job performed by a limited sequence robot. Presuming that their memory capacity is sufficient, they are also capable of more sophisticated jobs such as palletizing, stacking, spot welding and the like.

Playback robots — with continuous path control

There are, of course, applications in manufacturing industry where it is necessary to control not only the start and finish points of each robotized step, but also the path traced by the robot hand as it travels between these two extremes. A good example of this requirement is provided by seam welding, where a robot is asked to wield a welding gun, and move it along some complex contour at the correct speed to produce a strong and neat weld. One way of looking at this problem is to regard continuous path control as a logical extension of point-to-point control. It is feasible to provide a robot with a memory that is sufficiently large to allow path control that is, to all intents and purposes, continuous.

Alternately, the continuous path robot may be taught in real time. The operator takes hold of the robot by its hand, and leads it through the motions that it is going to have to perform by itself. The operator tries to copy the speed of travel required. During this teaching process, the robot has to record the movement and hand attitudes continuously, or approximately continuously, in its memory. This is achieved by giving the robot an internal timing system, which for example, could be synchronized with the 60 Hz main supply frequency. Using this time reference, the robot's movements are sampled at the rate of 60 times each second, with the results being committed to memory. Even at this sampling speed, a large amount of data has to be accumulated in the memory. Consequently, the storage systems for continuous path robots of this type often are magnetic tape units.

To increase the operational usefulness of continuous path robots, provision is usually made for the playback speed of operation to be different from the teaching speed. This is a great help to the operator, because in some applications the actual speed of travel is very slow, and the operator finds it easier to teach the robot at a faster speed, where he can make a smoother run, free from handshake. The converse is obviously also true. For higher speed continuous path programs as in applying a bead of sealant, it may be preferable to program at a lower than playback speed.

Arm geometry

A robot must be able to reach work pieces and tools. This requires a combination of an arm and a wrist subassembly, plus a hand, commonly called an end effector. The robot's sphere of influence is based upon the volume into which the robot's arm can deliver the wrist subassembly. A variety of geometric configurations have been studied and tried and their relative kinematic capabilities appraised. So far, robot manufacturers have selected one or more of the following:

○ Cartesian coordinates
○ Cylindrical coordinates
○ Polar coordinates
○ Revolute coordinates

Sketches of some typical embodiments are shown in Figure 2.6.

Evidently, each of these configurations offers a different shape to its sphere of influence, the total volume of which depends upon arm link lengths. For different applications, different configurations may be appropriate. A revolute arm might be best for reaching into a tub, while a cylindrical arm might be best suited to a straight thrust between the dies of a punch press.

In every case the arm carries a wrist assembly to orient its end

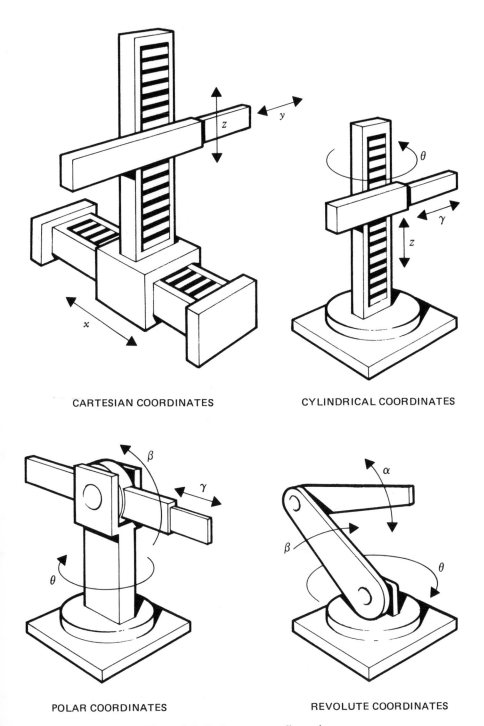

CARTESIAN COORDINATES   CYLINDRICAL COORDINATES

POLAR COORDINATES   REVOLUTE COORDINATES

Figure 2.6 *Robot arm configurations*

effector as demanded by workpiece placement. Commonly, the wrist provides three articulations that offer motions labeled pitch, yaw and roll — an obvious analogy with aircraft terminology. Figure 2.7 is a typical execution.

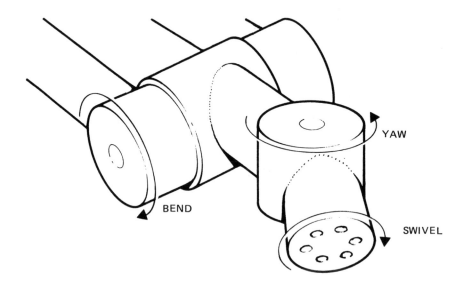

Figure 2.7 *Typical wrist articulations*

It may be noted that any of the arm coordinate systems requires three articulations to deliver the wrist assembly anywhere in the sphere of influence. It then requires three more articulations in the wrist for universal orientation of the end effector.

Quite often, robots are able to cope with job assignments without employing a full set of six articulations. This arises out of some symmetry in either the workpiece or the work place layout. For example, to move a bowling ball around in the sphere of influence requires only three articulations, because a ball is always oriented, irrespective of a gripper's orientation. More frequently, parts have one axis of symmetry (i.e. cylindrical) and this allows the robot arm to degenerate to five articulations.

Actually, five articulations are quite often adequate when the workplace is arranged to reduce part-manipulation needs. This happens, for example, when the beds of machine tools are all located parallel to one axis of a cartesian coordinate robot or on a radius of base rotation for cylindrical, polar or revolute robot arms.

Some might say that this compromising of number of articulations is begging the question of robots versus special-purpose automation. The robot should be the universal solution, readily transferred to other applications. In this vein, reference should be made to the elegance of computer control of robot arms. Given a six articulation arm of any configuration, software can permit a program to be generated in cartesian coordinates irrespective of the choice of articulations. Indeed, the software can be powerful enough to think only in tool coordinates. That is, the programmer concerns himself only with the tool on the end of the robot arm. He can think in terms of the tool's frame of reference and computer subroutines automatically make the various articulations move so as to accomplish the desired tool manipulation.

With computer control, the robot's geometry tends to lose installation significance. The engineering execution becomes the abiding issue.

Drive systems

A drive system is required for each robot articulation. In addition to driving the arm, the hand and the wrist, most types of grippers also need a drive mechanism for the functions of holding and releasing. Robot drives can be electrical, pneumatic, hydraulic or some combination of these.

Pneumatic systems are found in about 30% of robots, although they are confined mainly to the limited sequence devices. Pneumatic drives have the merit of being cheaper than other methods, and their inherent reliability means that maintenance costs can be kept down (although other types have also become reliable through progressive development). Since machine shops typically have compressed air lines available throughout their working areas, this makes the pneumatically driven robot a more familiar tool to shop personnel. Unfortunately, the system does not make for easy control of either speed or position, two essential ingredients for any successful robot.

Electromechanical drives are used in some 20% of robots. Typical forms are servomotors, stepping motors, pulse motors, linear solenoid and rotational solenoids. But the most popular form of drive system is hydraulic, because hydraulic cylinders and motors are compact, and allow high levels of force and power, together with accurate control. A more detailed comparison of the advantages and disadvantages between electrical and hydraulic drives is worth consideration.

Hydraulic versus electric drives

Hydraulic drives can be divided into actuators and motors. Actuators can be linear piston actuators or rotary vane actuators. Among the hydraulic motors there is a choice of piston, gear, vane and ball configurations. The choice is determined by several factors, such as the application, whether the motion required is linear or rotary, performance, requirement of lock-up, cost, reliability, and so on. The best choice is generally the simplest device that will do the job satisfactorily. Of all the hydraulic drives, the piston actuator is simple, very reliable, and the least expensive.

If better dynamic performance is necessary, the choice lies between making the piston actuators greatly oversized, or switching to hydraulic motors. Motors usually provide a more efficient way of using energy to achieve a better performance, but they are more expensive. Not only is the cost of hydraulic motors several times that of the actuators, but they also need auxiliary devices such as gearing or ball screws to complete the system. Whichever system is chosen, an electro-hydraulic servo valve is required.

Most electrical drives of robots are today in the form of rotational motors. Like the hydraulic motors, they need either gearing or ball screws to provide a complete actuation system. In the electrical system, the electro-hydraulic servo valve is replaced by a servo power amplifier.

It is seen that while electrical and hydraulic motor drives are approximately equal in complexity, hydraulic actuators are considerably simpler. The only drawback to the actuator solution is its lower stiffness. The question is, therefore, are motor drives for robots necessary? More often than not, the answer is no. Compare with machine tools that have natural structural frequencies in excess of 50 Hz — essential for them if they are to be chatter-free and effective chip producers.

Robots also need stiff structures and drives to be effective manufacturing tools, but on a different level. Few, if any, robots have a resonant frequency above 10 Hz. Machine tools are slow moving in relation to robots. The robot cannot afford fat to achieve stiffness. It needs rigid, but light, structures that can be adequately actuated by a piston cylinder rather than a bridge builder's dream moved by a screw.

The motor driven robot will have a much higher maintenance cost than the simpler actuator driven robots, not only because of the many more and costlier components, but because of localized wear in gears and ball screw by fretting corrosion during active servoing.

Fifteen to twenty years ago some machine tool builders switched from hydraulic to electrical drives. The reason behind the switch was believed to be reliability and the leakage problems of hydraulics. Since then some very significant improvements have been made in filters and seals, and with improvements in these areas a hydraulic servo system is certainly not second in reliability to a DC servomotor system. Oil leaks in hydraulic drives generally develop progressively. They are noticeable, they can often be repaired when convenient and they rarely cause unscheduled downtime. Nevertheless, they are annoying and can cause housekeeping problems. Danger of fire in some applications may require the use of phosphate ester or water-glycol types of fluid in place of conventional petroleum based hydraulic fluid.

In paint spraying and other applications the environment may present an explosion hazard and the robot must either be explosion proof or intrinsically safe so as not to ignite the combustible environment. Here the hydraulically driven robot has a great advantage over the electrical model since the electric energy from feedback devices and the energy to drive servo valves can be small enough not to ignite the explosive fuel-air mixture.

The last claimed advantage for hydraulics is that this power method lends itself to robot applications because energy can easily be stored in an accumulator and released when a burst of robot activity is called for. The 2000 series Unimate momentarily requires as much as 60 GPM which is supplied by a 17 GPM pump operating part time loading an accumulator. Because there are no convenient means to store electric energy, the designers of electrically driven robots tend to underpower the drives. To obtain the necessary dynamic performance they often use too high gear-ratios. The result is a snappy robot for small moves, but an embarassingly slow machine for large transfer moves.

In all of these discussions the question of size has so far been ignored. The cost of only some of the components is affected by size. The cost of a servo valve does not change much by size but the cost of an electric servo power amplifier will change greatly. The same goes for gears and other machined parts. A large, powerful robot normally works in more noisy and less pleasant surroundings than its smaller counterpart. A smaller robot will have an easier environment to contend with and probably a softer job. Also, because of its location, it will be treated better. All of these factors make it possible to design the small robot with a lesser safety factor than that required for a larger robot. As a result, the cost advantage of a hydraulically driven robot diminishes with size and when we talk about cost, we should consider total cost, including installation, maintenance and other operational expenses. All of that taken into consideration builds a very strong case for the electrical drive for the small robot.

The exact crossover point between hydraulic and electric drives may vary with robot configuration and the robot's intended use. Similar observations have been made by designers in other fields. Every drive has its pros and cons and will eventually find its proper place. In industrial handtools pneumatic motors are displacing electrical motors.

In some designs the proper choice was made at the start. We have electrically driven sewing machines and hydraulically operated backhoes, not the other way around.

Dynamic performance and accuracy

This discourse might very well have been entitled Dynamic Performance *versus* Accuracy, because these two qualities seem to be mutually exclusive. Closing out error of a servo to high accuracy is done at sacrifice of speed. While not as elegant a concept, it does appear to be as inexorable as the Heisenberg Uncertainty Principle, which, in particle physics, postulates that one cannot at the same time absolutely determine both the velocity and the position of a particle.

The analogy ought not to be overexercised. With diligent expenditure of money, ingenuity, and all the tools of modern servo theory, one can press the dynamic performance-accuracy impasse to levels well beyond that achieved by a simple proportional feedback servo.

A robot's speed can usually be evaluated in dollars (economic analysis in chapter 7 quantifies this for speeds 20% slower and 20% faster than a human operator). The marginal expenditure in gaining speed at no loss of accuracy is almost always cost effective.

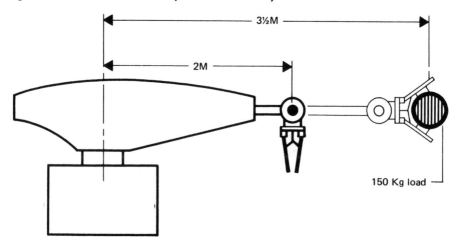

Figure 2.8 *Diagram of robot arm performance*
From fully retracted and unloaded to fully extended and carrying a 150 kg. load the moment of inertia changes from 70 KgMsec² to 230 KgMsec², a factor of over 3/1.

The problem can be quantified with reference to Figure 2.8. Consider a robot arm that has a retracted hand position of 2 meters and an extended hand position of 3½ meters. Consider also that this arm might carry a load of 150 kilograms, and that the arm should go from position-to-position, with or without load, at any extension and without overshoot. For the configuration of Figure 2.8, the variation in moment-of-inertia is from 70 KgMsec² when tucked in and unloaded to 230 KgMsec² when fully extended and loaded. To achieve a critically damped servo with position repeatability of 0.5 mm under all operating conditions is no mean chore. Note that 0.5 mm resolution for an arm with 300° of

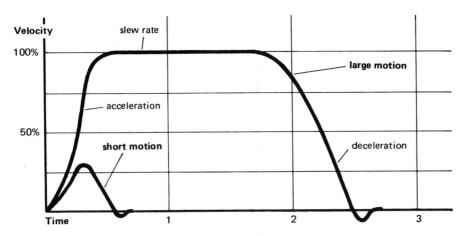

Figure 2.9 *Typical velocity traces for long and short arm motions*

rotation requires position encoding to an accuracy of 1 part in 33,000 or 2^{15}.

The foregoing deals only with a major robot arm articulation. In a full arm the interactions among the various articulations complicate both dynamic performance and accuracy. For example a robot arm designed to achieve an individual articulation natural frequency of 50 Hz, degenerates to an overall 17 Hz in a 6 articulation arm.

It is not the intention of this book to advise on robot design. Yet the manufacturing engineer should appreciate what a roboticist has to cope with in achieving requisite speed and accuracy. This may help protect him from 'specmanship'. The block diagram of Figure 2.10 functionally describes the key elements of a single articulation servo system with all of the 'bells and whistles' including velocity and acceleration feedback and inter-articulation bias signals.

To return to the issue of 'specmanship', it is common for robots to be offered with abbreviated specifications that list the slew rates of each articulation and the repeatability of each articulation. But, what is really needed is block point of time to go from position to position and net accuracy of all articulations in consort. Figure 2.9 shows two typical velocity traces for a short arm motion and for a large arm motion. It is evident that slew rate is no measure of elapsed time in making a motion, particularly a short motion in which slew rate may not be attained at all.

In estimating the time to complete a task (without actually simulating the entire process) the interface with the work place complicates the process. Paths to avoid obstacles add program steps. Some steps must be very precise, calling for closing out to zero error before the program advances. Other steps may be corners in a motion path which can be

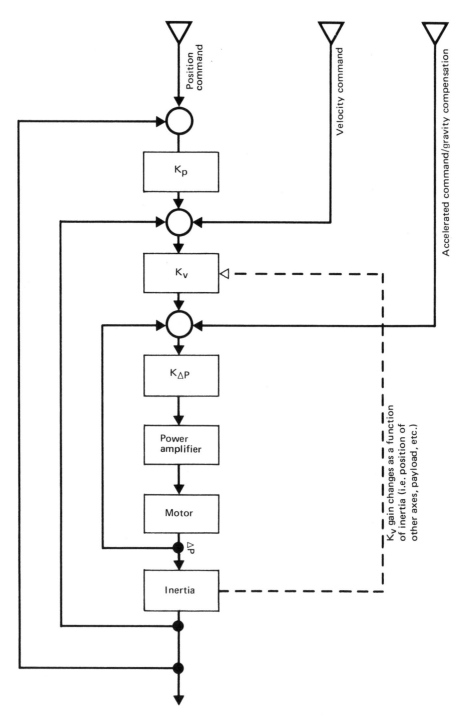

Figure 2.10 *Elements of a single articulation servo system*

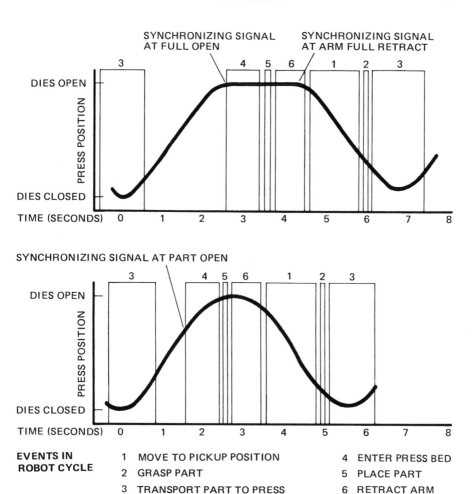

Figure 2.11 *The interlock system to reduce cycletime*

passed through on the fly. The use of interlock switches may introduce transport lags.

Simple programs often permit using a rule of thumb. For a 2000 series Unimate, if one allows 0.8 second for each motion taught, short steps as well as long, a time for program completion can be estimated quite closely. However, if a program is complex, as in spot welding a car body, there are too many variables to permit the use of such methods. In fact Unimation Inc. has developed an 18 page treatise to aid in forecasting program time for a multiplicity of robots spot welding an auto body. Factors considered are weld gun inertia, weld gun operating time, metal thickness, proximity of spots to one another, etc.

Sometimes program time is truly critical, such as when a robot is serving heavy, expensive capital equipment. If the production rate is paced by the robot rather than the capital equipment, the project is no longer viable because of loss in through-put.

Optimizing such a program may involve a range of tricks-of-the-trade. A typical application might be press-to-press transfer of automotive sheet-metal parts. A line of presses runs at a gross production rate of up to 700 parts per hour. At this rate a robot must make a complete transfer and return for the next pickup in 5.16 seconds. With presses on center-to-center distances of 6 meters, this is a demanding transfer speed. To meet this 700 per hour rate, a robot was modified by increasing the capacity of both hydraulic supply and servo valves. Acceleration and deceleration times were reduced at some sacrifice in damping and accuracy. This was compensated for by providing die nests with leads or strike bars. Finally, interlocks were refined so that the robot could make approaches and departures during the rise and fall of the presses' moving platens. Figure 2.11 shows how the time is shortened by tight interlocks that do not wait for press cycle completion. This strategy is not possible with human operators because of safety considerations.

Speed and accuracy! How is it that the human can do so well? The fully extended human arm has an unloaded natural frequency of only 3 Hz and the human's blind accuracy is much poorer than a robot's. What the human has is superb sensory aids. In motion the arm is guided by eyesight and proprioceptive muscle structure whose feedback mechanisms are not fully understood. Upon interactive contact, tactile data is added to the human's sensory richness. It may be that some measure of rudimentary vision and tactile sensing will ease the robot's servo demands when these qualities become commercially available. But, for the time being, the trade-off between dynamic performance and accuracy is a paramount design consideration. How a robot performs on this score is worthy of careful appraisal by the prospective user.

Chapter 3

End effectors: hands, grippers, pickups and tools

End effectors are at the business end of the robot. These are the moving components which have to grasp, lift and manipulate workpieces without causing any damage, and without letting go. Compared with human hands those of the robot are but clumsy travesties. They have fewer articulations and they are without any sense of feeling or touch. But, they can be designed to withstand high temperatures so that they are able to work with parts that are red hot. They are better at dealing with objects with sharp edges, or covered with corrosive substances, or which would simply be too heavy for human hands to grasp.

Being less adaptable than human hands, robot hands have to be chosen or designed specially for a particular industrial application. Whereas the robots themselves have earned the reputation of being general purpose automation, the hands are not quite so flexible and may have to be included along with the special tooling requirements of the job. In practice, this is not likely to prove a monkey wrench in the economic works. One type of hand is usually going to be suitable for a wide range of different jobs at a particular work station. Only when the robot has to be redeployed elsewhere to work on an entirely different process is it likely that the hand tooling has to be changed. And, compared with overall plant and machinery costs, hands come relatively cheap.

Methods of grasping

There are many ways of grasping or otherwise handling a job, depending to a large extent on the nature of the material being processed. Options include:
- Mechanical grippers.
- Hooking on to a part.
- Lifting and transferring a part on a thin platform or spatula.
- Scooping or ladling.
- Electromagnets.
- Vacuum cups.
- Sticky fingers, using adhesives.
- Quick disconnect bayonet sockets.

Some examples of appropriate methods are:
- O Forgings — normally handled by massive steel hands.
- O Thin metal sheets — vacuum cups and magnets are preferable in this case.
- O Powders, granular solids, liquids and molten metals — ladles or scoops.
- O Fabrics and similar flimsy material — vacuum cups, adhesives, and electrostatic devices all offer possible solutions. Usually much ingenuity is necessary.
- O Spot welding — weld gun permanently bolted to the robot wrist or exchangeable by means of bayonet socket.

Mechanical grippers

The following are the main factors in determining how grippers should grasp, and how hard.

1 The first and obvious rule is that the surface which the industrial robot's hand is to grasp must be reachable. As an example, it should not be hidden in a chuck.
2 Consider the tolerance of the surface we grasp and its influence on the accuracy in placing a part. If the machined portion of a cast part is to be inserted into a chuck — and the robot must grasp the cast surface — the opening in the chuck must be larger than the eccentricity between the cast and the machined surfaces.
3 The hand and fingers must be able to accommodate the change in dimension of a part that may occur between the part loading and the part unloading operations.
4 Consider how delicate surfaces are to be grasped and whether they may be distorted or scratched.
5 Select the larger if there is a choice of grasping a part on either of two different dimensions. Normally, this will assure better control in positioning the part.
6 Fingers should have either resilient pads or self-aligning jaws that will conform to the part to be picked up.

The reason for self-aligning jaws is to ensure that each jaw contacts the parts on two spots. If each jaw contacted the part on only one spot, the part could pivot between the jaws.

How hard the robot must grasp the part depends on the weight of the part, the friction between the part and the fingers (vacuum cups or magnet) how fast the robot is to move and the relation between the direction of movement to the fingers' position on the part. The worst case is when the acceleration forces are parallel to the contact surface of the fingers. Then friction alone has to hold the part.

A robot at normal full speed may, during acceleration and deceleration, very well exert forces on a part of about 2g (twice the earth's gravity). The following relationships are of interest:

1 A part transferred by a robot in the horizontal plane will exert a force on the hand tooling of twice the weight of the part.
2 If the part is lifted, it will exert a force three times its weight, 1g due to the earth's gravity and 2g due to acceleration upwards made by the robot.

The amount of friction which exists between the part and the fingers of the robot must also enter the picture. Consider the following example:

A weight of 25 pounds is to be lifted by a robot. The gravitational forces are parallel to the contact surfaces of the fingers and tend to pull the weight out of the hand. If the friction coefficient is 0.15, how hard must the robot grasp the part? Include a reasonable safety factor in the solution. The equation for this situation is:

clamping force x friction coefficient   =   tangential force
                                        =   weight of the part
                                            x g-loading.

This reduces to:

clamping force x 0.15 = 25 x 3
i.e. clamping force   = 500 lbs

With a safety factor of two, the clamping force should therefore be 1000 pounds.

If the center of gravity of the part is outside the line between the two jaws, a moment due to acceleration forces will tend to pivot the part. To prevent pivoting the product of the clamping force, the spread between contact points and friction must be greater than the moment. An example of the method of calculating grasping force is given in Figure 3.1.

A part weighs 25 lbs and has a center of gravity 15″ off the point where it is grasped. The friction coefficient is 0.15 and the spread between the contact point on each jaw pad is 3″. How hard must the robot grasp to have a safety factor of 2 on its hold of the part?

It is necessary to design for the highest force which will occur at the point where slippage between fingers and the part will first arise, if the clamping force is not high enough.

Figure 3.1 *Example showing calculation of grasping force*

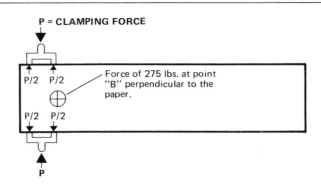

clamping force x friction coefficient (μ)
$\geqslant$ force at point "B"
2 $(P/2 \times \mu) = B$
2 $(P/2 \times 0.15) = 275$
   $P = 1830$ lbs
with a safety factor of 2, the required
clamping force is 3660 lbs.

This is a very large clamping force for a 25 lb part, which indicates that the design is not efficient. The force can be reduced in two ways: first, try to grasp the part closer to CG or make the pads longer (6" instead of 3" will reduce the clamping force from 3660 to 1340 lbs).

The situation may be represented thus:

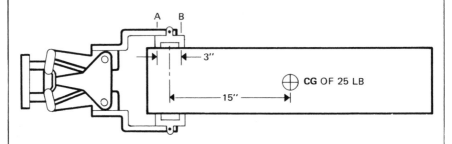

This simplifies into the following force diagram:

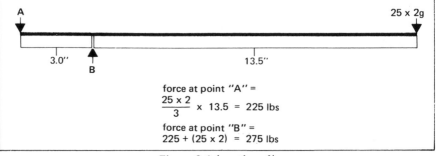

force at point "A" =
$$\frac{25 \times 2}{3} \times 13.5 = 225 \text{ lbs}$$

force at point "B" =
$225 + (25 \times 2) = 275$ lbs

Figure 3.1 (continued)

Some indication of the very wide variety of mechanical grippers which have been designed to meet different robot applications can be gained from Figure 3.2. The selection is by no means representative of all that is available or possible. As previously mentioned, grippers are the one area of robotization where specialized design of tooling is often necessary though it is seldom expensive.

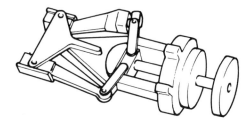

Standard hand
This is an inexpensive and all-purpose hand that will accept a virtually infinite variety of custom fingers. Fingers are tailored to the parts to be manipulated or moved. The parts should be of moderate weight. Simple linkages provide both the finger action and the force multi-plication needed to grip the object sufficiently tightly. At the completion of finger closure, the fingers exert their maximum clamping force on the part.

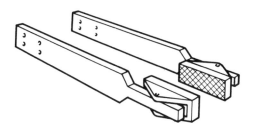

Fingers self-aligning
Self-aligning pads for fingers are valuable for assuring a secure grip on a flat-sided part. 'Cocking' of the part is highly unlikely when these pads are employed.

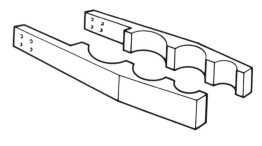

Fingers for grasping different size parts
A particular finger design need not be restricted to parts within a limited range of sizes. Perhaps the fingers can be equipped with extended pads having several cavities for parts of differing sizes and shapes, or for parts that change shape during processing. Then, the industrial robot is pre-programmed to position the hand so that the proper cavity will match the location of the part.

Figure 3.2 *Examples of mechanical grippers*

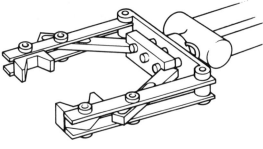

Cam-operated hand

Heavy weights or bulky objects are handled easily by the cam-operated hand. More expensive than the standard hand, the cam-operated hand is designed to hold the part so that its center of gravity (CG) is kept very close to the 'wrist' of the hand. The short distance between the CG and wrist minimizes the twisting tendency of a heavy or bulky object. To achieve this 'close coupling' of hand and part, there is a sacrifice: a specific cam-operated hand design will accommodate only a very narrow range of object sizes.

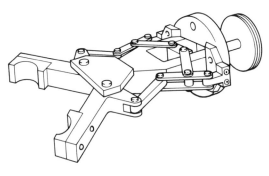

Wide-opening hand

When the part to be picked up is not always to be found in a constant orientation or at the same site, a wide-opening hand may be recommended. As it closes, this hand will sweep the inexactly located part into its grasp.

If the part to be grasped is always precisely positioned for pick-up, the wide-opening hand can shorten the time needed to reach for the part. The hand can travel the shortest path to the part and skip the extra step of making its final approach to the part from one specific direction. The hand develops low force when open and maximum force when closed. It is for parts of moderate weight.

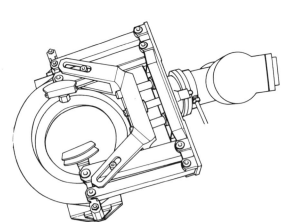

Cam-operated hand with inside and outside jaws

Assume that a part is re-oriented between the time when the part is placed in a machine and when it is removed. This special hand is one of those which will deal with this problem. When the part is oriented as shown, the hand can grasp it on the OD by employing the outer self-aligning pads. If the part is turned over, the inner pads will grasp the ID.

A similar principle applies when the grasped surface of a part is changed significantly between the time when it is placed in a machine and the time when it is removed. A special hand can be designed to deal with most changes in ID, OD, or other dimension.

Figure 3.2 (continued)

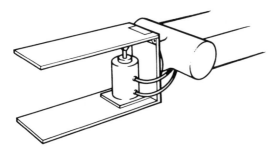

Special hand with one movable jaw
A hand with single-acting jaw should be considered when there is any access underneath a part, as when it is on a rack. Where this hand can be applied, it will scoop up a part quite quickly. Simplicity of the design makes this one of the most economical hands.

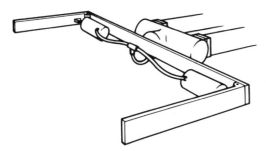

Special hand for cartons
The dual-jaw hand will open wide to grasp inexactly located objects of light weight. Lifting and placement of cardboard cartons is an application. Actuators and jaws can be re-mounted in any of several positions on the fixed back plate, making it practical for the same dual-jaw hand to move large cartons on one day and smaller cartons the next.

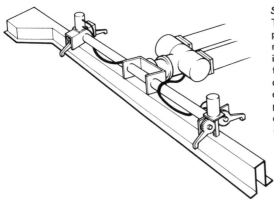

Special hand with modular gripper
This special hand, with pair of pneumatic actuators, is one of the many special hand designs for industrial robots. It would be suitable for parts of light weight. Lifting capacity is dependent upon friction developed by the fingers, but heavier parts could be handled if the fingers could secure a more positive purchase — as under a flange or lip.

Figure 3.2 (continued)

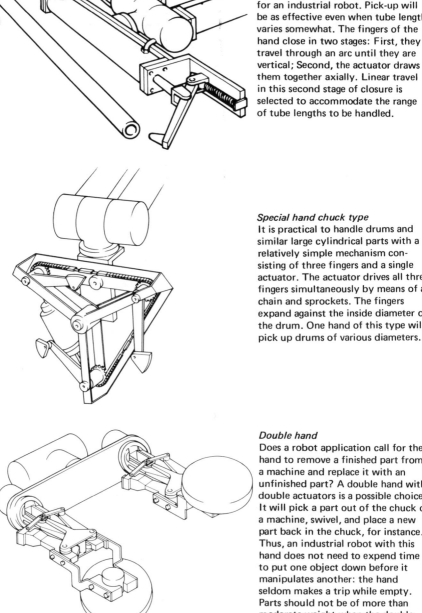

Special hand for glass tubes
Secure grasping of relatively short tubes is the forte of this special hand for an industrial robot. Pick-up will be as effective even when tube length varies somewhat. The fingers of the hand close in two stages: First, they travel through an arc until they are vertical; Second, the actuator draws them together axially. Linear travel in this second stage of closure is selected to accommodate the range of tube lengths to be handled.

Special hand chuck type
It is practical to handle drums and similar large cylindrical parts with a relatively simple mechanism consisting of three fingers and a single actuator. The actuator drives all three fingers simultaneously by means of a chain and sprockets. The fingers expand against the inside diameter of the drum. One hand of this type will pick up drums of various diameters.

Double hand
Does a robot application call for the hand to remove a finished part from a machine and replace it with an unfinished part? A double hand with double actuators is a possible choice. It will pick a part out of the chuck of a machine, swivel, and place a new part back in the chuck, for instance. Thus, an industrial robot with this hand does not need to expend time to put one object down before it manipulates another: the hand seldom makes a trip while empty. Parts should not be of more than moderate weight when the double hand is used.

Figure 3.2 (continued)

Vacuum systems

Vacuum cups

Vacuum cups are normally made of an elastic material that conforms and forms a seal to the surface of the part to be handled. If the part is elastic, then of course the cup can be made of a hard material. The shape of the cups is mostly what the name implies — cup-shaped.

There are other configurations that differ in principle to the usual cup. Some cups, or vacuum pads, are made of cellular material through which the air is drawn. They have the advantage of working on a rough and porous surface; e.g., a common brick, because each cell constitutes its own little vacuum cup and if one fails to make a seal, it is paired up with neighboring cells and they together form a larger cup.

The holding force of a vacuum cup is the effective area multiplied by the different of pressure between the outside and inside of the cup.

The effective area of a cup is often not the geometric area, because the cup often deforms when vacuum is applied. If the bottom of the cup touches the object to be lifted, the effective area is reduced correspondingly.

To get the best utilization of a cup, the largest possible vacuum or pressure differential should be used. In most cases, which we will deal with later, it is better to use a larger cup and a lower vacuum to obtain a faster system.

The vacuum will not form until the cup has sealed on the part; therefore, to get speed out of a vacuum system, it is advantageous to mount the cups on spring-loaded stems and have the robot programmed so that the cup touches the part long before the arm reaches its final pick-up position. This will eliminate a large portion of the deceleration time from the cycle.

Springloading of the cups will also compensate for any variation in the height or level of the part. If there are any variations between the parts to be handled, like distorted sheet stock, it can often be compensated for by putting the cups on ball joints, as well as springloaded stems.

For sliding of the parts, the same rule applies as for fingers: the force multiplied by the friction coefficient between the cups and the material. If oily sheet stock is being picked up, the coefficient will not be simply the friction between rubber and metal; this is normally very low and, in most cases, the cup will not break through the oil film. In such cases' viscous friction rather than Coulomb friction is present, which means that the sheets will always slide sideways to some extent when exposed to a force.

The life of vacuum cups is quite good, especially in relation to their price. Polyurethane cups seem to have a longer life than those made from natural or synthetic rubber. Vacuum cups are catalog items and there is a selection to choose from in both configurations and sizes.

The number of cups to be used in a design depends on such factors as: weight of the load, size of cups available, location of the center of gravity and the support needed to handle large flimsy parts.

Vacuum pump versus *venturi*

To create a vacuum a choice exists between two devices, the vacuum pump or the venturi. A vacuum pump is either a piston or vane-type pump driven by an electric motor. The venturi is a device where vacuum is created by having a secondary high energy stream of flow impinge on the primary flow actually converting pressure into vacuum. The advantages of a pump are:
- Able to create a high vacuum
- Low cost of operation
- Relatively silent

Disadvantages:
- High initial cost
- Requires a more complex system: vacuum tank and blow-off valve

The advantages of a venturi:
- Low initial cost
- Does not normally need blow-off valve or vacuum tank
- High reliability

Disadvantages:
- Very noisy
- High cost of operation

The venturi system differs from the pump system in that it is not controlled by a valve in the vacuum line but rather by control of the high pressure air to the venturi.

By this control mode, the venturi is working only when vacuum is desired. Consequently, the size of the venturi has to be of full size and cannot utilize a low-duty cycle to charge and draw peak loads from an accumulator.

Since there is no valve in the vacuum line, the response time is not limited by it. Instead, the response is a function of the size of the venturi, and in the cases where the venturi is turned on after contact, by the time delay in the pressure line.

One simple way to make an estimate of the response is to establish the lowest pressure at which the vacuum cups can pick up the load and where this vacuum intersects the proper supply pressure line, read off the corresponding flow. Dividing the volume of the vacuum cups and the lines by this flow will yield the time it takes to evacuate the cups. This estimate is conservative since the venturi has a higher flow at lower vacuum.

Special considerations

While attention has been paid to the engagement of the parts to the cups, little has been mentioned about the disengagement which is of equal importance.

To release a part fast, a blow-off system is required. A typical arrangement of such a system is shown in the schematic of the pump-vacuum system.

In studies it has been found that the limiting factor of the speed of response is the size of the valves. Direct operated solenoid valves seldom have larger flow areas than 1/8th inch holes. For higher capacities of high speed valves, it is necessary to use a pilot operated valve. The ultimate system would then consist of a pilot-operated solenoid valve and plumbing sized accordingly.

Some typical vacuum pick-up systems are illustrated in Figure 3.3.

Magnetic pickups

Parts handling for various processing operations can be accomplished in several ways but if the parts are of ferrous content, magnetic handling should be one of the methods to receive consideration.

Magnets can be scientifically designed and made in numerous shapes and sizes to perform various tasks. A ferrous object placed within the range of a magnet will itself become magnetized, and will then have its own North and South poles which will be attracted to the parent or larger magnet *in proportion to its mass.*

Magnets fall into two principal categories, namely permanent and electro. Either of these types can be adapted within reasonable limits to handle parts having various shapes and often it is possible to handle several different shapes with the same magnet.

Electro magnets are well suited for remote control as well as for moderately high speed pick up and release of parts. A source of D.C. power is required in connection with control equipment which should be selected for the specific application. To assist in releasing parts without hesitation, an item known as a 'drop controller' is incorporated in the circuit. Basically, it is a multi-function switch through which power is supplied to the magnet and as it interrupts the power supply, it reverses the polarity and supplies power at a reduced voltage for a short duration before completely disconnecting the magnet from the line. This reverse polarity tends to cancel any residual magnetism in the part to make sure that it will release instantaneously.

Permanent magnets do not require a power source for operation which makes them well adapted for hazardous atmospheres that require explosion proof electrical equipment. They do, however, require a means of separating material from the magnet. To accomplish this, a stripper device may be employed or if the part is positioned and

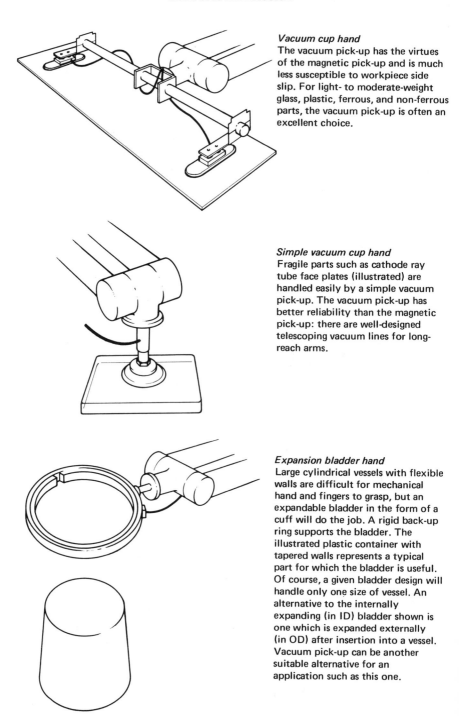

Vacuum cup hand
The vacuum pick-up has the virtues of the magnetic pick-up and is much less susceptible to workpiece side slip. For light- to moderate-weight glass, plastic, ferrous, and non-ferrous parts, the vacuum pick-up is often an excellent choice.

Simple vacuum cup hand
Fragile parts such as cathode ray tube face plates (illustrated) are handled easily by a simple vacuum pick-up. The vacuum pick-up has better reliability than the magnetic pick-up: there are well-designed telescoping vacuum lines for long-reach arms.

Expansion bladder hand
Large cylindrical vessels with flexible walls are difficult for mechanical hand and fingers to grasp, but an expandable bladder in the form of a cuff will do the job. A rigid back-up ring supports the bladder. The illustrated plastic container with tapered walls represents a typical part for which the bladder is useful. Of course, a given bladder design will handle only one size of vessel. An alternative to the internally expanding (in ID) bladder shown is one which is expanded externally (in OD) after insertion into a vessel. Vacuum pick-up can be another suitable alternative for an application such as this one.

Figure 3.3 *Some typical vacuum pickup systems*

clamped, welded or otherwise secured the magnet can be pulled from the part. The permanent magnet can be designed to produce extremely shallow magnetic penetration, a feature that is valuable when, for example, it is necessary to remove single thin ferrous metal sheets from a stack. In fact, standard designs are available that will lift single sheets as thin as .031 inch.

Another version of a permanent magnet which can be used with sheets is the sheet 'fanner' or separator. This device separates sheets in a pile so a magnetic 'hand' or a gripper can pick up individual pieces. Magnetic induction of the sheets with like polarity causes them to repel each other and to tend to rise in mid-air. As each sheet is taken away the others rise to higher positions.

Regardless of whether the magnet employed is permanent or electro, there are several matters that must be considered before a proper selection can be made. The following are of importance.

1 SHAPE OF PART

Parts having a large flat contact surface are 'naturals' for magnetic handling. Other shapes can be handled, but more compensation must then be made. For example, round pieces tend to roll but this can be prevented by providing pole plates with contoured or irregular surfaces. Any part having relatively high mass in relation to the area presented for magnetic contact will require a stronger magnet to project enough magnetic flux lines into the material to permit lifting.

2 WEIGHT

It is obvious that the lifting capability of the magnet must be great enough to handle the heaviest part to be manipulated. Conditions are seldom perfect, and just as a crane cable or sling must be selected to have some reserve capacity, a magnet must be given the same consideration.

3 TEMPERATURE

Electromagnets of standard design will handle materials having temperatures up to 140°F. Modified designs will accommodate temperatures up to 300°F and special designs can be made for even higher temperatures although cost then becomes more of a determining factor. Most permanent magnets are fully effective if material temperature does not exceed 200°F though others have been designed for handling parts up to 900°F.

In all cases, frequency and length of contact with the hot part and the length determine most of the operating limits.

4 SURFACE CONDITION

A smooth, flat, dry, clean surface is ideally suited for magnetic parts handling. Irregular or curved surfaces will affect holding power but after compensation has been made for the irregularity, performance is then predictable. However, rust, mill scale, oil, pits and sand can individually or collectively affect holding power in an unpredictable manner so it is advantageous to minimize these conditions as far as possible.

5 POSITION TO BE HANDLED

Lifting, transferring or otherwise handling parts in such a manner that they are directly beneath the magnet is the most efficient way of handling. But having the magnet face in a vertical plane with parts cantilevered only uses a magnet at 25% or less of its maximum potential because the material tends to slide rather than pull away from the face. Some parts held in this position might possess a shape that would create a bending moment which would tend to break the part away from the magnet.

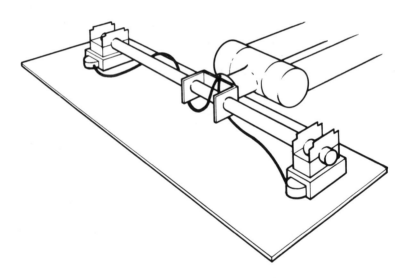

These pickups are good for use on flat surfaces, such as ferrous sheets or plates, and will deal with objects of several sizes. Weight of the part should be no more than moderate so that side slippage is avoided. Positioning for pickup does not need to be precise and 'grasping' is instantaneous, both time savers.

Figure 3.4 *Typical electro magnet pickup for use with flat surfaces*

As previously stated, the electrical power required for an electro magnet must be D.C. This can be supplied by batteries, engine or motor driven generator sets or rectified A.C. Batteries offer the greatest portability but are the most limited in capacity. Many plants have generators to supply D.C. for other operations so often there is a ready source available. If it is necessary to resort to rectified A.C., this poses few problems since rectifier design is constantly improving and the present costs of such a system are not prohibitive.

A typical electromagnetic pickup designed for use with flat surfaces is illustrated in Figure 3.4.

Tools

With various grasping and pickup devices, robots clumsily imitate what a human operator might do. Sometimes the human is directed to pick up a tool and use it continuously. When a robot takes over such a task the tool might just as well be fastened to the robot's extremity permanently. Or, if the robot has two or more tools to choose among, then quick disconnect selection of tools may be in order.

Apart from peculiar mounting characteristics, tools fastened to robot wrists are likely to be given the same capabilities they would have had if they were manually manipulated. Therefore, the concept needs only to be documented with examples of tools affixed to robot wrists.

A range of such tools is illustrated in Figure 3.5.

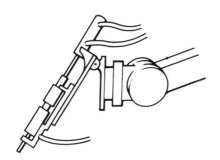

Stud-welding head
Equipping an industrial robot with a stud-welding head is also practical. Studs are fed to the head from a tubular feeder suspended from overhead.

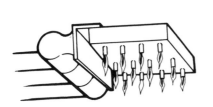

Heating torch
The industrial robot can also manipulate a heating torch to bake out foundry molds by playing the torch over the surface, letting the flame linger where more heat input is needed. Fuel is saved because heat is applied directly, and the bakeout is faster than it would be if the molds were conveyed through a gas-fired oven.

Figure 3.5 *Examples of tools fastened to robot wrists*

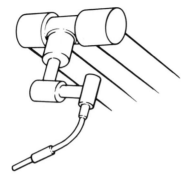

Inert gas arc welding torch
Arc welding with a robot-held torch is another application in which an industrial robot can take over from a man. The welds can be single- or multiple-pass. The most effective use is for running simple-curved and compound-curved joints, as well as running multiple short welds at different angles and on various planes.

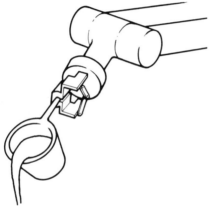

Ladle
Ladling hot materials such as molten metal is a hot and hazardous job for which industrial robots are well-suited. In piston casting, permanent mold die casting, and related applications, the robot can be programmed to scoop up and transfer the molten metal from the pot to the mold, and then do the pouring. In cases where dross will form, dipping techniques will often keep it out of the mold.

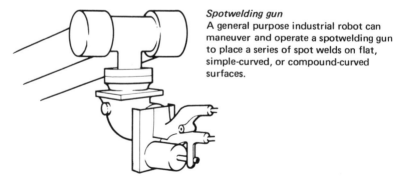

Spotwelding gun
A general purpose industrial robot can maneuver and operate a spotwelding gun to place a series of spot welds on flat, simple-curved, or compound-curved surfaces.

Figure 3.5 (continued)

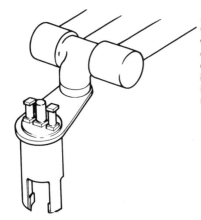

Pneumatic nut-runners, drills and impact wrenches
General purpose industrial robots are especially well suited for performing nut-running and similar operations in hazardous environments. Drilling and countersinking with the aid of a positioning guide is another application. Mechanical guides will increase the locating accuracy of the robot and also help shorten positioning time.

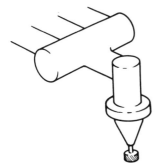

Routers, sanders and grinders
A routing head, grinder, belt sander, or disc sander can be mounted readily on the wrist of an industrial robot. Thus equipped, the robot can rout workpiece edges, remove flash from plastic parts, and do rough snagging of castings.

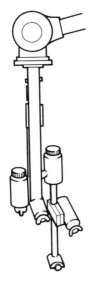

Spray gun
Ability of the industrial robot to do multipass spraying with controlled velocity fits it for automated application of primers, paints, and ceramic or glass frits, as well as application of masking agents used before plating. For short or medium-length production runs, the industrial robot would often be a better choice than a special-purpose setup requiring a lengthy change-over procedure for each different part. Also, the robot can spray parts with compound curvatures and multiple surfaces.

Figure 3.5 (continued)

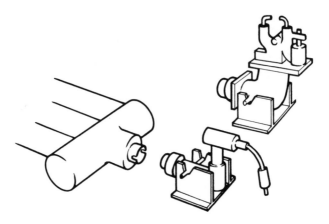

Tool changing
A single industrial robot can also handle several tools sequentially, with an automatic tool-changing operation programmed into the robot's memory. The tools can be of different types or sizes, permitting multiple operations on the same workpiece. To remove a tool, the robot lowers the tool into a cradle that retains the snap-in tool as the robot pulls its wrist away. The process is reversed to pick up another tool.

Figure 3.5 (continued)

Chapter 4

Matching robots to the workplace

Robotizing a process might mean anything from the purchase of a single robot to replace one man at an existing machine to the design and implementation of a complex manufacturing system using several robots, all controlled from a central computer. Even in the very simplest case, there is more to consider than choosing the best robot for the job, asking the existing man to step aside, and setting the robot to work in place of him. In this chapter some of the practical aspects of the robot-to-machine interface are examined. While it is soon found that special provisions must be made to compensate for the robot's inability to see and feel, there are many opportunities for cashing in on some of the non-human – even superhuman – properties of our mechanical imitators. Intelligent production engineering and system design should seek to explore ways and means for taking full advantage of the robot's capacity for working continuously, accurately, and reliably under hostile conditions – not forgetting that work can sometimes be arranged with one robot tending two or more machines in a work sequence that would quickly have a human completely worn out.

Part orientation

Whatever the size of the system, and whether there is just one robot or a whole battery of them, one of the things to get sorted out right from the start is the physical position and attitude of the workpiece at the front end of the work station. It would be no good at all, for instance, simply to pile a batch of parts in a random heap in a tub or tote box, dump the lot down, and expect the robot to be able to pick out the parts one by one in a sensible, repeatable gripper-to-part relationship, at least not with technology in its present state of development. All that the robot could do would be to grope blindly at the pile of parts, with a chance that one or more might be picked up, but with no chance at all that any part would ever be gripped in the position and attitude necessary for it to be fed to the next stage of the process.

The solution adopted must obviously be suited to each case, since it will depend on the particular manufacturing operation and the relative dispositions of all the related items of plant. Some discussion of this problem is undertaken in the case histories which form Part II of this book. Whatever the method, things have to be arranged so that the robot can be taught to pick up the first part of a batch correctly, after which all the other parts in the batch are presented to the robot hand at the same pickup point, in the same attitude, or at least in a series of known attitudes. Palletization often provides a suitable answer. Parts can be located on a spigot, or between guides. Sometimes parts can be arranged on a pallet in a grid pattern, and the robot must then be taught to recognize this pattern and pick up the parts in sequence until the pallet has been emptied. Discs can be set up in a neat vertical stack from which the robot plucks them one by one, using vacuum cups instead of gripper fingers. In molding, die casting, and similar processes part orientation is far less of a problem for the simple reason that each new part is made in exactly the same place, and in precisely the same attitude.

Part orientation is not just a matter of knowing where the workpiece is to be found when the robot picks it up for the first time. Consider, for example, the common arrangement for quenching die castings by dropping them into a bath of cold water. This may be an obvious and convenient way of cooling the castings when a man is operating the machine, but it is by no means so clever when a robot is doing the job. If the robot is expected to take the raw castings from the quench tank and load them into a trim press, it is going to look pretty silly fishing for them in the water where their position and orientation is anybody's guess. The answer in this case is to make the robot grasp the part, take it from the mold, dip it into the water tank, and *without letting go,* load it into the trim press. Although this means that the work has had to be adapted to recognize the robot's shortcomings, full advantage has been taken of the fact that robot hands can grip parts that would be too hot for a man to handle.

So far, the examples given have all been concerned with the loading or operation of machines. In other applications, such as the welding of car bodies, the work is brought to the robot on a conveyor, stopped while the robot does its job, after which the conveyor steps on to bring the next body into position. Part orientation is determined only by the accuracy with which each body is located on the conveyor system. In such stop-go conveyor arrangements, the conveyor speed is matched to the slowest operation along the line and each movement of the conveyor is triggered from a central control system. More complicated is the process where the work is offered to the robot on a conveyor which does not stop, but which causes the workpiece to be carried slowly past the robot station. Moreover, the speed of such conveyors may be variable, to suit progress achieved at other stations along the line. The complex problems which this creates for the robot are overcome with the aid of instrumentation which senses the conveyor speed and which can signal the exact position of the workpiece to the robot's own command system.

There is significant work underway to provide robots with some rudimentary sensory perception—visual and tactile. When these attributes become available there will be less insistence upon absolute preservation of orientation. For now, however, in any robotized manufacturing system the rules should be:

— Arrange for part orientation to be defined at the pickup point.
— Once the process has started, never allow part orientation to be lost.
— Never drop a part!

Interlocks and sequence control

In order to establish a good working relationship between the robot and its associated plant, interlocks and sensors have to be provided that replace the ears, eyes, nose and hands of the human worker. Such devices are needed to initiate each stage of the production cycle at the right time, and to prevent damaging or dangerous movements of any part of the robot or plant. Thus, the conveyor must not start up before the robot has removed a part clear of the delivery station. It is obviously desirable that a robot arm is not placed between the closing jaws of a press. Has the robot really removed all the casting from the die casting mold or are some broken pieces still trapped in there waiting to cause havoc when the next cycle starts? Production engineers starting up a new robotized work center have to weigh up all the normal requirements of the process, then consider possible malfunctions which need special interlocks or sensors to protect the hardware, and the entire control system must be designed to fit the robot into the workplace in a sensible, integrated fashion.

Fortunately, there is no shortage of mechanical, electrical and electronic devices that can be built into a total control system. Designing these controls and interlocks is within the competence of any process control engineer used to working with automated machinery. In robotics, the engineer finds a slightly different situation from designing the custom-built type of work station because he must find ways and means for fitting sensing devices, limit switches and the like, to standard machinery that was intended for operation by human hands. Such modifications are seldom difficult, but their introduction should not in any way prevent the general purpose machinery in the system from being redeployed elsewhere in the future or make it difficult for the machinery to revert to manual operation should the robot be out of service. Some suggestions follow.

Mechanically operated limit switches: clamped to machine slides, conveyors or to any other place where the position of a moving part is critical to starting or stopping the robot sequence.

Microswitches: useful in conjunction with end stops to act as limit switches or to sense the weight of parts stacked on a pallet. For example, a pallet can be arranged to sit on a spring-loaded platform, so that when the workpiece is in place on the pallet, the platform is depressed sufficiently to operate a strategically placed microswitch underneath.

Photoelectric devices: capable of sensing the presence of any object, provided that the object is opaque, when the object interrupts a beam of light.

Pressure switches: arranged to monitor the pressure of air lines or hydraulic feeds. For example, the pressure could be monitored at the cylinder of a fixture clamp, so that the robot could be signalled when the clamping pressure released and the workpiece was ready for extraction.

Vacuum switches: arranged so that if the robot is operating with a vacuum type pick up unit, the robot does not move from the pickup position until a vacuum is indicated.

Infrared detectors: capable of detecting the absence or presence of hot workpieces. These are particularly useful in such applications as die casting and forging. Infrared detectors can also be used to check that parts are at the correct temperature for the process.

Signals from other electronic control systems: a most important source of sequencing information. These sources might be NC machines, other robots, or a computer arranged as a master controller for the entire manufacturing system.

Since it is possible both to send and receive signals to and from the associated equipment and also to utilize this information at any desired point in the robot program, the robot now has the ability to accept complete control over any required sequence of operations. An additional advantage is that the sequence of operations may be varied automatically depending upon the information received from the associated equipment.

The alternative sequences required will vary from application to application and from the simple to the complex. In the majority of applications the variable sequence may be controlled by information given to the robot by simple limit switches, proximity detectors, etc.

Outline example of a sequence control problem

In a typical application the robot is required to interface with two incoming conveyors, two pallets, two safety doors and a reject position. The requirements are that the robot should pick up parts from conveyor A and load pallet A, pick up from conveyor B and load pallet B. After pickup from either conveyor, the parts are presented to a detector located at each conveyor. The detectors give a GO/NO GO signal. Should a NO GO (reject) signal be received, the parts from either conveyor are placed at the common reject position.

Since each pallet is loaded with 8 layers at 5 parts per layer, the robot memory must also have the capability to memorize how many parts and in what position the last part was loaded on each pallet.

Other requirements of this application are:

1 While parts are present at each conveyor the robot must alternate between conveyors.

2 If parts are present on only one conveyor or there are parts present on both conveyors but with a 'queue' indicated on one conveyor, then the robot must give priority to the most loaded conveyor. That is, either the conveyor with parts available or, alternatively, to clear the queue.

3 Should there be a queue on both conveyors the robot must alternate until such times as one queue is cleared and then revert to priority on the other conveyor.

4 Should the reject position become full and the robot has a reject part, then the robot must stop until space is available.

5 Having fully loaded either pallet the robot must revert to loading the remaining pallet. Should both pallets be fully loaded the robot must stop.

6 When an empty pallet replaces a fully loaded pallet the robot must automatically recognize the condition on each conveyor and revert to alternate or priority.

The sophisticated industrial robot with a large memory and with the ability to digest a range of inputs and dispense a range of outputs can cope with an almost bewildering spectrum of alternative actions — so long as all possibilities have been anticipated by the system designer.

Detailed analysis of setting up a sequence control system

This example is drawn from a real-life report* submitted in the automobile industry and presented here in its authentic form.

*This section, including Figures 4.1, 4.2, 4.3 and 4.4, has been extracted with slight modifications from Dennis W. Hanify and Jay V. Belcher, *Industrial robot analysis — working place studies* (Proceedings 5th International Industrial Robot Symposium, Sept, 22-24, 1975, Chicago, Ill.), published by the Society of Manufacturing Engineers, Dearborn, Mich., 1975, to whom due acknowledgement is expressed.

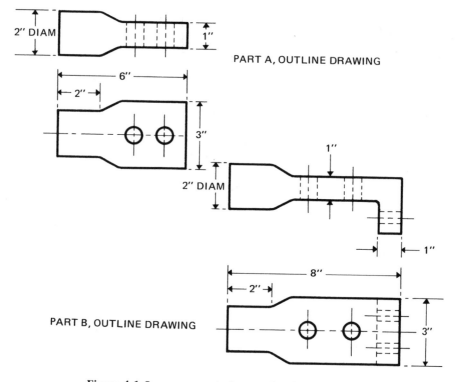

Figure 4.1 *Sequence control example: the workpieces*

Two different workpieces are to be manufactured at the workplace. They are very similar in shape and the machining operations are the same for both pieces. Workpiece outline drawings appear in Figure 4.1. Each workpiece weighs approximately 1.82 Kg. The material is an aluminum alloy forging. Workpiece feed positions are shown for each machine in Figure 4.2. Machining is done on the hub of each part.

MATERIALS HANDLING SEQUENCE
The previous machining operations are automated and at the completion of the machining cycle the part will be manually placed in a holding fixture, properly oriented and ready for the subsequent machining cycle. The sequence of this machining cycle is:
1 Part is removed from the holding fixture and moved to Machine A.
2 The part is then inserted into the clamping fixture on Machine A and the machining cycle started.
3 After the automatic machine cycle, the part is moved to Machine B.
4 Part is inserted into the clamping fixture of Machine B and the automatic machining cycle started.
5 After completion of the automatic cycle the part is moved to Machine C.

6 Part is inserted into clamping fixture of Machine C and the automatic cycle started.

7 After completion of the automatic cycle the part is moved to Machine D.

8 Part is inserted into the aligning and clamping fixture and the automatic cycle started.

9 After completion of the automatic cycle the part is moved to Machine E.

10 Part is inserted in the holding fixture and the machine cycle started.

11 After completion of the cycle the part is loaded in a tote bucket or rack for disposition.

Part positioning for inserting in the machine fixtures is also shown in Figure 4.2. A maximum positioning error of not greater than 1.6 mm is required. The machine fixtures have been designed to allow this tolerance and still maintain the machining accuracy required.

ANALYSIS OF THE MACHINING CYCLE

The production machines involved in this study are sufficiently automated to be used with an industrial robot. The fixtures used with these machines provide the required degree of automation for clamping and aligning the part.

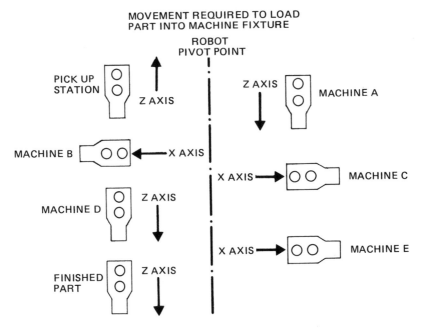

Figure 4.2 *Sequence control example: workpiece feed positions*

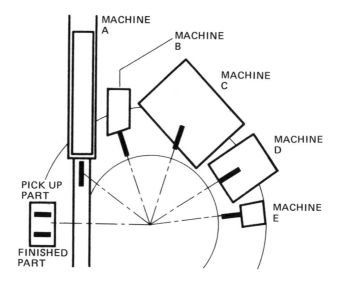

Figure 4.3 *Sequence control example: equipment layout*

Since the machine tools involved in this study are flexible in their placement, the machining sequence should be kept in mind and the machines located in a logical sequence of machining operations. A typical layout is shown in Figure 4.3. This type of layout permits the part to be transferred from one machining operation to the next in the correct sequence. The layout eliminates wasted robot movements and minimizes cycle time.

The maximum cycle time allowed for the five machining operations is 1 minute 45 seconds to 2 minutes. This period of time is dictated by the previous operation and by the requirement of 1200 parts per week, which is 30 parts per hour on a one-shift basis. The cycle times which include clamping, machining and unclamping for each machine are:

Machine A:	11 seconds
Machine B:	6 seconds
Machine C:	20 seconds
Machine D:	15 seconds
Machine E:	10 seconds

INTERLOCKS ANALYSIS

When interlocks are properly used and are of sufficient number, costly collisions, jam-ups, scrap parts, and costly damage to the robot and equipment can be prevented. The following list of interlocks represents the minimum number required in this example and can be expanded as desired or as the equipment permits.

Pickup station
Signal to robot that a part is present.

Machine A
Signal to robot that fixture clamp is open.
Signal to robot that fixture is empty.
Signal to robot that fixture clamp has closed.
Signal to robot that machine cycle is complete.

Machine B
Signal to robot that fixture clamp is open.
Signal to robot that fixture is empty.
Signal to robot that fixture clamp has closed.
Signal to robot that machine cycle is complete.

Machine C
Signal to robot that fixture clamp is open.
Signal to robot that fixture is empty.
Signal to robot that fixture clamp has closed.
Signal to robot that machine cycle is complete.

Machine D
Signal to robot that fixture clamp is open.
Signal to robot that fixture is empty.
Signal to robot that fixture clamp has closed.
Signal to robot that machine cycle is complete.

Machine E
Signal to robot that fixture clamp is open.
Signal to robot that fixture is empty.
Signal to robot that fixture clamp has closed.
Signal to robot that machine cycle is complete.

Signals from the robot to the machines are also important for automatic operation. These three signals are applicable to all machines: close fixture clamps; start machining cycle; and open fixture clamps.

GRIPPING REQUIREMENTS

The gripping technique required is a dual gripper design. This type of gripper permits handling two parts at a time so that a part may be removed from a machine and the new part inserted without excessive motions of the robot and a loss of cycle time. Rubber pads should be used on the fingers to give some compliance and protect the part finish.

In designing the gripper it must be kept in mind that two different parts must be handled and the same program should be used for both parts. This permits parts to be intermixed in the machining operation.

Workplace layout

Work configurations can be classified in the following four ways:

1 Arranging work around the robot
2 Bringing work to the robot
3 Work travels past the robot (a variant of 2)
4 Robot travels to work

Naturally each configuration is appropriate for different manufacturing operations or systems of work organization. One of the early decisions in the installation process is to establish an optimal working layout.

Arranging work around the robot

All early installations were of the first class, because this involved the least commitment and the least plant disruption for the oft-times skeptical pioneer user. In die casting, for example, the first tentative step was to put the robot in front of the already-installed die casting machine and let it extract and quench the casting. Since the robot had time on its hand, it was not a very bold step forward to bring in a trim press, put it in reach of the robot, and let the robot operate both machines. When, as is often the case, a second die casting machine is close at hand, it may be practical to have just one robot unloading two die casting machines, quenching trimming and stacking the output of both, an evident illustration of the 'surrounded by work' class of operation. See Figure 10.4 in Part 2 of this book.

In loading and unloading metal cutting machines, cutting times are often such that one robot can attend to a group of machines. A logical layout for the polar coordinate robot arm is to group the machine tools around the robot, within its sphere of influence. This 'surrounded by work' installation remains the most prevalent in the field. Such jobs as forging and trimming, press to press transfer, plastic molding and packaging, and investment casting are other examples of the class.

Work travels past the robot

The addition of computer control to an industrial robot produces tremendous flexibility. For example, the robot can be made to track a workpiece which is being carried on a conveyor, performing its task as the job passes by — see Figure 4.4. The versatility of such a system can be extended to cope with variations in the conveyor speed.

The following description* describes the systems by which line tracking with an industrial robot can be accomplished.

MOVING-BASE LINE TRACKING

With this method, the robot is mounted on some form of transport system, e.g. a rail and carriage system, which moves parallel to the line and at line speed. This method requires the installation of the transport system which may not be possible or economical. If multiple robot systems are set up adjacent to one another alongside a moving line,

*This section, including Figures 4.5 and 4.6, has been extracted in slightly modified form from the paper *Moving line applications with a computer controlled robot*, by Bryan L. Dawson, Applications Engineer, Cincinnati Milacron, SME Technical Paper MS 77-742, published by the Society of Manufacturing Engineers, Dearborn, Mich., 1977, with due acknowledgement to author and publisher.

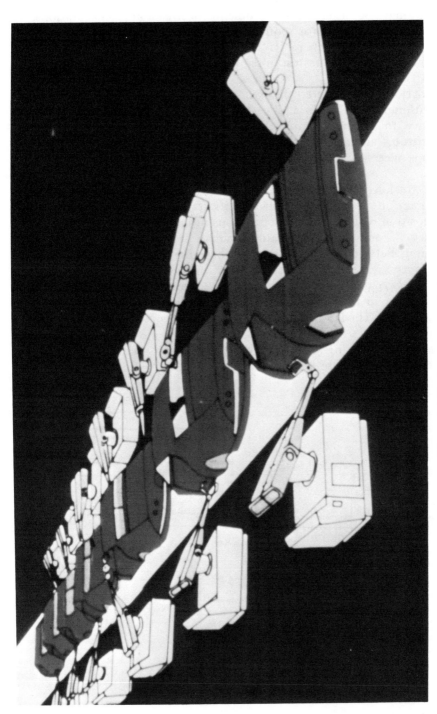

Figure 4.4 *Work comes to robot*

there may be interference problems between adjacent stations. A powerful drive system is required for each transport device in order to return the robot back to its starting point from the other end of its tracking range in the fastest possible time.

2. STATIONARY-BASE LINE TRACKING

In this method of line tracking, the robot is mounted in a fixed position relative to the line. Hence the name 'stationary-base'. This naturally constitutes an economical installation which requires less maintenance than is necessary with moving-base systems.

Full tracking capability of the robot allows it to perform its taught program on a part moving through its station, irrespective of the speed or position of that part. The positions of taught points, the orientation angles of end effectors around taught points and the velocities of motions between taught points will have the same values, *in relation to the part*, system of the computer controlled robot allows the full tracking capability to be easily implemented. Positions of taught points are stored in memory as coordinates in space and not as robot axis coordinates. The layout of the system is summarized in Figure 4.5.

During the teaching operation the part is moved to a convenient position in front of the robot and stopped. Points are taught as normal but each coordinate in the direction of the line is modified by an

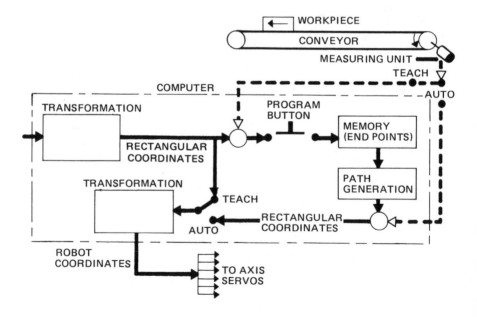

Figure 4.5 *Work travels past robot — diagram of tracking and control system*

amount equal to the current position sensor reading, prior to being stored in memory. Thus, the stored data are referenced to the start point of tracking. If it is desirable, for more convenient access, the part may be repositioned at any time during the teaching operation.

In the automatic mode of operation, the stored points are used to generate the desired paths, which are then modified by the current position sensor reading. In this way, the control, in effect, changes the coordinates of taught points in the tracking direction by an amount equivalent to the distance between the position of the part at which the point was taught and the position of the part where the point is replayed.

IMPLEMENTATION OF STATIONARY-BASE LINE TRACKING

The requirements for a typical robot installation to be used for a stationary-base line tracking application are:

1 A position sensor connected to the part of conveyor to indicate the position of the part. This sensor is electrically interfaced with the control.

2 A limit switch or other form of sensor which is actuated when the part is in a predefined position. This sensor signal, called 'Target In Range', indicates to the control to start to use the information provided by the position sensor to update the position of the part.

3 A series of limit switches or sensors which indicate to the control the style of part on which the robot is to operate. This permits the control to select the correct branch program for that part from its memory. These switch or sensor signals use a simple binary code to allow the control to select one of 15 different branch programs.

CONSIDERATIONS FOR STATIONARY-BASED TRACKING APPLICATIONS

If a sequence of operations to be performed on a stationary part is taught to a robot, the robot will replay the programmed points at the same positions, in space, at which they were taught. The points will always be within the range of the robot arm during replay because it is impossible to teach a point that is outside that range. However, when a robot is working on a continuously-moving part, taught points on the part that were within the range of the robot during the teaching operation may, due to a variety of circumstances, be outside that range during replay. Points that were taught with the part at one end of the range could be replayed with the part at the other end of the range. Hence, because the robot will not be replaying programs with modified paths between modified programmed points, there are certain considerations to be taken into account in the planning and programming of moving line tracking applications with a stationary-base robot. These are discussed in the following text.

Tracking window. The diagram in Figure 4.6 illustrates the robot's large tracking range, when used in tracking applications in which the Y axis of the robot is set parallel to the moving line. As the diagram indicates, there are many parameters that influence the length of working range of the robot in the direction parallel to the moving line. This working range of the robot parallel to the line is termed the 'tracking window'. The height of the part on the conveyor, the distance of the robot from the conveyor and the length and configuration of the end effector all play a part in determining the tracking window. Therefore, every tracking application must be considered separately in order that the robot is positioned correctly, relative to the conveyor, to ensure the optimum tracking window.

Once the tracking window for a given sequence of operations has been established, it is entered into the memory of the control. The tracking window basically defines in memory the two limits in the tracking direction beyond which the robot will not attempt to reach. More than one tracking window may be defined for different segments of a tracking operation.

Abort branches and utility branches are available in software. They ensure that, when the robot is working with a moving line, logical decisions and actions are made by the control to take corrective action in response the occurrence of random but foreseeable events. As with non-tracking applications, other interface signals between the robot and the peripheral equipment are easily implemented to ensure that corrective action is taken by the robot in response to other occurrences.

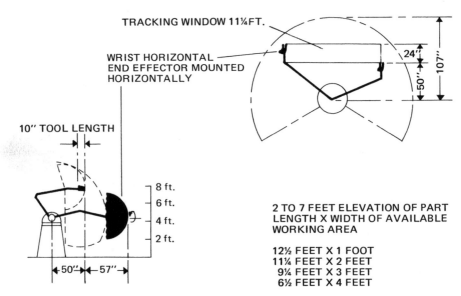

Figure 4.6 *Work travels past robot — examples of tracking windows*

Robot travels to work

When machining cycles are particularly long, a robot can be mounted on a track to enable it to travel among more machines than can conveniently be grouped around a stationary robot. Figure 4.7 is a photograph of a track mounted robot that handles eleven different machine tools. In this example, a buffer station is carried with the robot for parts in intermediate stages of completion.

Figure 4.7 *Robot travels to work — track mounted robot serving 11 machine tools*

In Figure 4.8 a robot is shown which travels overhead to service eight NC lathes. This installation is controlled by a central computer, which instructs the lathes and the robot. The control room contains a library of machining programs for the lathes, as well as for all the possible loading and unloading programs used by the robot. The central computer also choreographs the travels of the robot up and down the line to minimize individual lathe downtime. The line is 200 ft long. Figure 4.9 is a schematic representation of the system.

The system is more fully described in Part II, chapter 19.

Figure 4.8 *Robot travels to work — overhead robot serving eight NC lathes*

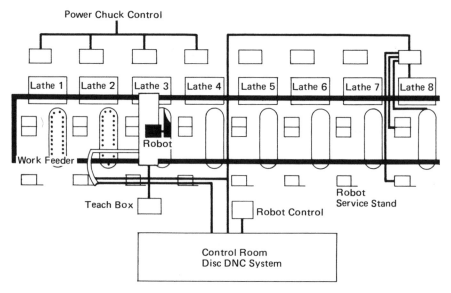

Figure 4.9 *Robot travels to work — diagram of overhead robot system portrayed in Figure 4.8*

Chapter 5

Reliability, maintenance and safety

Experience over millions of robot operating hours in manufacturing industry has proved that robots can be both reliable and safe. Their reliability has been proven by demonstrating the ability to work for hour after hour, day after day, in hostile conditions. They do this with only the rare breakdown. When they do break down, as often as not their downtime is short, because diagnostic routines enable the maintenance workers to get them back on stream without delay. On a parallel with their reliability performance, robots have been proved safe: accidents involving injury to human beings are exceedingly unusual. Serious injury has never occurred. These impressive performance records in reliability and safety have been brought about by attention to design and by commonsense application of robots in use.

Environmental factors in robot systems

When equipment is designed for a specific purpose, such as a military application, it is usually possible to specify a set of environmental conditions surrounding the operation of that equipment. These conditions then form part of the design specification, and they play a large part in the choice of individual components, and in the layout and construction of the final hardware. Such is not the case for industrial robots. There is no single set of environmental conditions which covers all possible industrial possibilities. Nevertheless, after many millions of hours of industrial robot experience, it has been possible to come up with a design and quality test procedure that seems to have conquered the hazards presented by most known industrial environments. Test and design methods are, to some extent, empirical, and so will continue to evolve as more and more experience is gained.

Figure 5.1 lists the primary environmental factors that have to be considered in robot design. Because these conditions are described qualitatively, there may be merit in expanding upon this tabulation with even more qualitative discussion to give the would-be robot designer a 'gut feeling' for the job at hand.

```
1  Ambient temperature: up to 120°F without cooling air
2  Radiant heating: source temperature up to 2000°F
3  Shock: excursions up to ½ inch, repetitions to 2/second
4  Electrical noise: line drop-outs, motor starting transients; RF heating
5  Liquid sprays: water and other coolants, often corrosive
6  Fumes & vapors: process chemicals, steam cleaning
7  Particulate matter: sand, metallic dust, hot slag
8  Fire & explosion risk: open flame, explosive gas & vapor mixtures
```

Figure 5.1 *Hazards in the industrial environment*

HEAT

Ordinarily, a human worker is not required to function continuously in an ambient temperature over $120°F$ and therefore this is a reasonable maximum standard for an industrial robot. Both the human operator and the industrial robot are afforded cooling air if the workplace temperature exceeds $120°F$. In some instances particular attention must be paid to radiant heating where the worker is the target of open furnaces, lehrs and hot parts in process. Radiation shields are sometimes used and a robot may expect to be provided with a curtain quench for its extremities.

SHOCK AND VIBRATION

There are not too many instances when an individual robot must endure severe vibration conditions. It is usually lugged to massive floor members and vibration from associated equipment is minimal. On the other hand, shock can be severe. Some hammer forge operations develop shock so severe that it can be felt in offices 300 yards away.

ELECTRICAL NOISE AND INTERFERENCE

One of the most frustrating environmental conditions to plague a robot designer is electrical noise. Designers have been unable to create a noise standard which would enable us to extrapolate in-house testing to noise immunity in the field. Any new design is put into the field in operations which we have found to be particularly 'dirty' as regards electrical noise.

An electrical line dropout that might cause a computation error in a computer, means only a burst of 'garbage' data output. For an industrial robot it might mean physical action damaging to the robot or to the equipment with which it is associated. Noise insensitivity is crucial and without a clear definition of the noise environment, design becomes an iterative process cycling back and forth between the field and the laboratory.

LIQUID SPRAYS, GASES AND HARMFUL PARTICLES

There are lots of things that land on or diffuse through industrial robots in factories. These are often the same things that are designated as health hazards to human operators. The 'black lung' human debility has its counterpart in the susceptibly designed industrial robot. Some examples will help make the point:

○ In investment casting, the atmosphere is heavily contaminated with alcohol-ammonia fumes which are highly injurious to any open switch contact. The same operation also includes particulate contamination and gear trains and sliding bearings require absolute protection.

○ In one foundry application carbonized-silica particles are continuously in the atmosphere and fall to the floor at the rate of ¼ inch per day. The material is extremely abrasive, and with any moisture at all, corrosive to electronics.

○ Heat treatment processes often involve combination of high temperature and high humidity. This may be compounded by the salt solution which is used in heat treatment and which can build up on the industrial robot.

○ In some forging operations, die lubricant is applied copiously and the process impels the lubricant onto all surrounding equipment. This water-suspended graphite-based material builds up on the equipment and steam cleaning may be necessary daily. Steam cleaning itself is one of an industrial robot's environmental hazards.

○ Sparks fly continuously in resistance welding set ups and flying metal particles will bond to open metal surfaces.

○ A robot does not stand aside when molten metal is shot into a casting machine and therefore it may be exposed to slag spurting out between die parting lines. The slag is hot, it must be endured and the robot must occasionally suffer hammer and chisel removal of the build up.

RISK OF FIRE OR EXPLOSION

A man is inherently non-flammable and when admonished not to smoke he is non-igniting. When a robot stands in at a job where there is a continuing open flame, there must be protection against leakage of flammable servo oil that might cause a serious fire. In a job where an explosive atmosphere is created by the volatile carriers of paint, for instance, the robot must not have any design element capable of creating a spark.

Examples of hazardous situations

The following cases illustrate robots being subjected to some of the risks generated in typical industrial applications.

○ In Figure 5.2, a robot services a die casting machine, suffering the perils of its stance at the die parting line. Note the protective

Figure 5.2 *Hazardous situation: robot services die casting machine*

baffle on its boom cover and the skirt below.

O Figure 5.3 shows a robot transferring billets in and out of a rotary furnace for heating just prior to a forging operation.

O In high energy forging, the billet is formed in one pass. Figure 5.4 shows the robot re-entering the press bed to extract the formed part.

O The chips created in machining are a hazard to robots. A boot around extension rods plus tight seals on wrist gearing are essential to long life in the application shown in Figure 5.5.

O Figure 5.6 shows the robot on a spot welding line. The robot must endure sparks, oil leaks and cooling water spray.

Designing robots for industrial environments

In designing industrial robots the primary influence is the nature of the jobs to be done. Once manipulative power, sphere of influence, speed, strength and memory capacity have been established, environmental conditions can be brought to bear upon the design.

Some examples of design concessions to the environment follow:

○ Considering some of the hot and hostile places that a robot's hand must enter, it is desirable to eliminate all electrics and servos from robot extremities.

○ In many applications it is entirely proper to package the robot as a self-contained entity, but there is an advantage to a design in which the electronics may be mounted separately. In extreme

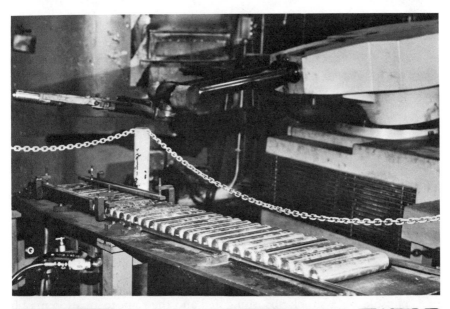

Figure 5.3 *Hazardous situation: robot transferring billets in and out of rotary furnace*

Figure 5.4 *Hazardous situation: robot re-entering press bed*

shock conditions it is convenient to be able to mount the control console on a shock absorbing pad and the remote location may be necessary to ensure life in a corrosive atmosphere. Going further, if the power supply of the robot can also be separated from the robot's arm then the arm may be introduced all alone into explosive atmospheres such as paint rooms.

○ With hot metal often flying about, it is important that all of the robot's skin be of non-flammable construction.

○ Where robot articulations are exposed in rotating or sliding joints, the joints should be booted to protect against abrasive dust collection.

○ A part answer to fire hazard conditions is provided by non-flammable fluids for lubrication and hydraulics. Such are available as an option because there is a significant cost disadvantage to their introduction.

○ If air is particularly dirty, air cooling may not be practical and the option to use water cooling should be made available. In any event, all cooling air should first go through filtration and enter enclosures to provide positive internal pressure.

○ Dust and dirt seem to be able to infiltrate the tightest crevices and therefore drive trains should use hardened gears and be pressurized to exclude contamination.

○ Robot logic design should be heavily protected from power line spikes and noise pickup entering through any of the robot's communication links with surrounding equipment.

Figure 5.5 *Hazardous situation: robot protected from machining chips*

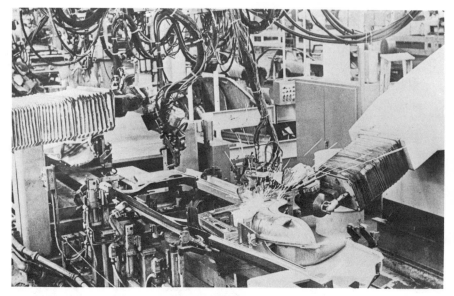

Figure 5.6 *Hazardous situation: robot subjected to sparks, oil leaks and water spray on spot welding line*

Reliability targets

There are two concepts to be considered in a discussion of reliability: one is Mean Time Between Failure (MTBF), and the other is 'downtime'. Clearly, there is a relationship between the time interval between failures and the total amount of downtime; but the correlation is not linear because there is the additional variable of Mean Time To Repair (MTTR). Thus, an otherwise satisfactory MTBF could result in unacceptable downtime if the time to repair is excessive.

An industrial robot is usually working in conjunction with another piece of productive machinery. When the industrial robot is down, and there is no provision for manual backup, the production machinery is also down. Industrial experience indicates that for most applications, uptime must exceed 97% to satisfy the users of industrial robots. This rule of thumb is somewhat dependent upon the specific application. In an operation such as die casting, which inherently includes a lot of downtime, the process is not as sensitive to robot downtime. In a glass manufacturing plant, which is very akin to a continuous process, there may be need for uptime of 99.5% and if this is not attainable, provision must be made for backup labor which might be supplied by a backup relief human crew or by spare robots.

If the design goal is a downtime of no more than 2%, and if we then conjecture that the robot manufacturer can always offer next-day service, and that all robots are working on a two-shift basis, the most likely downtime per incident becomes 8 hours. If the downtime per incident is 8 hours then a 400-hour Mean Time Between Failure must be the standard to hold overall downtime under 2%.

Theoretical reliability assessment

Given a rough target to aim at for downtime, it is interesting to consider how the theoretical reliability performance of a robot is evaluated. In the case of the Unimate robot, which has a long design history, the design standard was arrived at against a reliability study covering all the system components. This study was exhaustive, and the theoretical contribution of every component towards a total system failure rate was taken into account. This work was undertaken by an independent organization, Bird Engineering Research Associates, Inc., which specializes in reliability assessment. The Bird Report to Unimation Inc. concluded that Unimation's system design justified achieving an MTBF of 500 hours without incurring prohibitive costs in component design, system design, manufacturing, and quality control procedures.

The complete analysis was, of course, voluminous, but an appreciation of the method can be gleaned from an overview. To predict the failure rates for all of the components in the Unimate, Bird relied upon notebooks which were prepared under U.S. Government contract to aid prediction of reliability of space vehicle systems. For electronic components, the Rome Air Development Center Notebook, TR-67-108, was used and for mechanical hydraulic components Bird used the U.S. Navy's Failure Rate Data (FARADA) Notebook. Both of these references were cross-correlated with other similar data banks.

Figure 5.7 is a tabulation of reliability feasibility for electronic/ electrical elements. It will be noted in this table that the expected MTBF for parts only is 1800 hours. Because of the contribution of other (non-parts) the estimated MTBF drops to 1217 hours. The non-part failure rate is a system failure rate due to tolerance buildup, critical interface tolerances, customer abuse, unforeseen environmental problems, etc. This type of failure occurs in analog systems at a rate proportional to the complexity of the system and for the Unimate this was established by the consultants by comparing complexity to similar systems for which massive data exists. Evidently, the upper limit for an MTBF would be the 1800 hours due to parts only. This upper limit could be approached only at high dollar cost and therefore if the more modest 1217 hour goal were not sufficient, it would be necessary to examine other alternatives including going back to the original design concept.

Figure 5.8 is a table which integrates the electronic/electrical and mechanical/hydraulic failure rates to predict an overall attainable

Component or Element	Part Failure Rates $(x10^{-6})$
Common Group	
Power Supply	54
Shift Registers	103
Memory	112
Relay Tree	125
Control Panel	11
Memory Sequence Control	16
Home Options (3 x 60)	11
Scanner	3
Comparator	14
Sequence Control	7
Operate External	18
Wait External	15
Counter-Demod (Common)	4
Subtotal	493
Servo Loops (5)	
Counter-Demod	4
Servo Power Amp	16
Servo Switch & Dir. Store	13
Encoder Electronics	29
Subtotal	62
Total Electronic/Electrical Failure Rate $(x\,10^{-6})$:	
Parts Only	555
Other (Non-parts)	267
Overall System	822
Electronic/Electrical MTBF (hrs):	
Parts Only	1800
Other (Non-parts)	3745
Overall System	1217

Figure 5.7 *Reliability of electronic/electrical elements used in Unimate 2000 Series design*

MTBF for the Unimate system.

Once the reliability feasibility has been established, the hard work really begins. Aiming for an overall MTBF of 400 hours, Unimation Inc. set up a management system designed to bring individual components up to standard and assure statistically that the system, as shipped, will meet the overall goal. Figure 5.9 shows the reliability control points in the Unimate life cycle.

Since field experience is crucial to determining true reliability, the entire process of building toward this reliability includes placing machines in the field and feeding back the results of this experience into the reliability control system. In the case of the 2000 Series Unimate, the opening experience produced an MTBF of 145 hours and over the ensuing years of production, this MTBF was slowly brought up to 415 hours.

Failure classification	Failure rate (x 10^{-6})	MTBF (hours)
Part failures only:		
Electronic/Electrical	555	1800
Mechanical/Hydraulic	673	1485
Non-part failures:		
Electronic/Electrical	267	3745
Mechanical/Hydraulic	475	2100
System failures:		
Parts only	1228	815
Non-tolerance	742	1350
Combined	1970	508

Estimated reliability feasibility,
Unimate 2000: MTBF = 500 Hours

Figure 5.8 *Unimate system reliability estimate*

Long range outlook for reliability

The succeeding generations of industrial robots will become ever more sophisticated and that might be expected to portend a much reduced reliability potential. Fortunately, the reliability of solid state electronics continues to improve, and this counters the natural inverse correlation of reliability with complexity.

Nevertheless, with proliferating use of industrial robots in the factories and with more intimate interlocks between robots and factory information systems, it will become more and more difficult to maintain satisfactory uptime in the complete manufacturing process. It seems that the touchstone to minimizing downtime in the fully automated factory will be found in diagnostic monitoring systems which will pinpoint trouble spots instantaneously. Even in the face of deterioration in MTBF, really significant improvements in reaction time to a failure incident will result overall in acceptable factory uptime.

Maintenance needs and economics

The planning of robot maintenance can therefore draw upon a body of statistical information about the reliability of these machines according to hours worked and environmental conditions.

Field experience of reliability

In monitoring the continuing reliability of their robots, Unimation selects a sample of machines, chosen for their activity and spectrum of applications. Furthermore, the Unimate sample includes only those machines which are serviced by Unimation Inc personnel, to ensure consistency of data and to exclude those breakdowns which might result from a manufacturer's own maintenance shortcomings. In Unimation's reliability analysis it is assumed that all components have a constant failure rate. This simplification does not take into account wear and time-dependent degradation of performance. Data has to be watched for evidence of deterioration in the MTBF, which would suggest that a major overhaul is necessary. Unimation's experience indicates that deterioration of reliability is very much dependent upon the application, but seems to become significant after operating periods of between 8,000 and 15,000 hours. That is equivalent to something between four and seven man-shift years.

The discussion so far has been centred on MTBF, but it is well to consider also downtime. Four hundred hours was chosen by Unimation as a reasonable MTBF when speculating on acceptable downtime. Assuming that all field service is done by their personnel, that all their robots are working on a two shift basis, and that they can always offer next-day service, then the most likely downtime per incident that Unimation could expect would be 8 hours. This is 2% of the operating time, given an MTBF of 400 hours.

Of course, the situation is more complex. There is a variety of opportunities for minimizing the impact of downtime. For one thing, if the user has his own trained maintenance personnel, he should be able to react to a failure without having to await the arrival of the robot supplier's personnel. So too, even where training of a customer's personnel has been minimal, it is often possible to get a machine back on stream by telephoning the robot manufacturer. The user describes the breakdown symptoms, and the expert robotics engineer can attempt a remote diagnosis and suggest a prescribed course of treatment for the sick robot.

Maintenance economics

Unimation Inc. usually recommends that there be an eventual potential for at least three industrial robots at any single plant location. A trained maintenance man will not be effectively utilized until there is a minimum of three robots for which he is responsible. Exceptions, however, include cases in which risk-control concerns were dominant or in which the productivity gains achievable with one or two robots would more than off-set the expense of a specially trained employee.

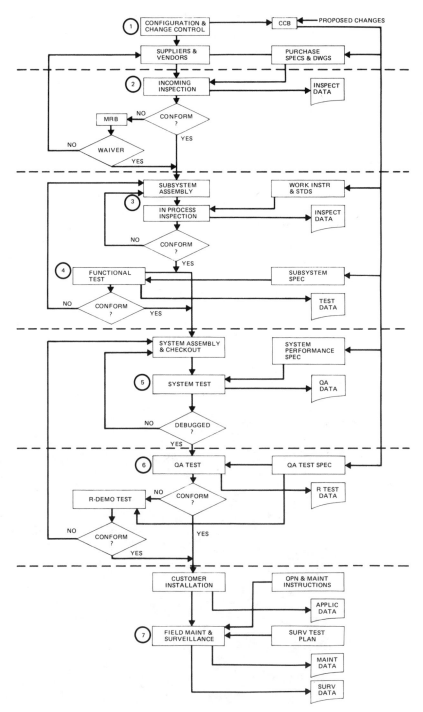

Figure 5.9 *Reliability control points in the Unimate life cycle*

Manufacturers of general-purpose industrial robots operate customer training programs or can arrange service contracts.

High cost inflation makes it unwise to try and give any absolute indication of maintenance costs. A more permanent approach is to compare the maintenance costs for robots with their cost of acquisition. This is the approach commonly used for factory automation in the automotive industry, where the annual maintenance costs are characteristically some 10% of the plant acquisition costs. For a Unimate robot working two shifts (4,000 hours) the annual maintenance works out at about 11% of average acquisition cost.

Under ideal circumstances, the user who has a multiplicity of robots in his plant can organize his own trained maintenance personnel. He is able to invest in the robot manufacturer's diagnostic test instruments (where these are offered for sale) and he can afford to build up a sensible stock of replacement parts, commensurate with the size of the robot population in his plant. Under such conditions the MTBF-downtime-maintenance pattern can be optimized.

Figure 5.10 shows a portion of the Unimate line in the Lordstown plant of General Motors. This plant hardly ever calls upon Unimation Inc. and all of its service is performed by blue-collar workers. Remarkably, with 26 machines on the line, downtime of the entire line amounts to only 6 minutes per shift. With an MTBF of 400 hours and 26 machines in the system, we can expect a robot failure incident every 15 hours. If the line were to be shut down upon every robot failure, it

Figure 5.10 *The Unimate line at General Motors' Lordstown plant*

would not be possible to hold downtime to 6 minutes per shift. But GM shuts the line down only as a last resort; normal procedure is to lock a failed robot out of the system and assign a human operator to pick up the missed weld spots at the end of the line. The serviceman on duty uses his knowledge of the equipment and the diagnostic test instrument to determine what corrective action is necessary. After he performs the service work, he puts the robot back on stream and signals the relief operator that the line is going once again at full strength.

Auto manufacturers with multiple industrial robots usually have spare robots available for production-line operations, such as spot welding. A fork-lift truck side-lines any robot needing repairs, and replaces it with a spare. The program is quickly extracted from the memory of the defective robot and written into the memory of the spare robot, or a duplicate program on file can be transferred into the spare robot's memory, typically within three and one-half minutes.

Safety levels and precautions

Accident prevention has become one of the professions, with people acting as watchdogs to ensure that things are built to specified, safe, standards. Other authorities watch over the workplace, enacting legislation which attempts to prevent dangerous practices, and try to save people from the risks offered by particular machinery, or simply to do everything possible to prevent men from hurting themselves through their own carelessness. None of these measures can guarantee a plant free from accident. Wherever people work there will always be some risk. All that can be done is to bring this risk down by identifying the major threats, by predicting what could go wrong, and by taking all sensible precautions before the worst happens.

And so, with robots as with any other new technology, we have to ask the questions 'Are they safe?' 'What are the special risks?' 'What special precautions are necessary?' Sensible consideration of such questions soon starts to reveal some very reassuring answers. Isaac Asimov's first law of robotics was part of the robot designer's brief — that a robot shall never injure a human being or, through inactivity, allow a human being to come to harm. This may only be an abstract ideal, but it is a principle that has been borne out dramatically through millions of operating hours. Before going on to consider the precautions needed to prevent human accidents with robots, it has to be declared that robots have made a positive contribution to safety at the workplace, by removing man from the vicinity of other equipment that really does pose a serious threat to life and limb. Undoubtedly, serious accidents at power presses, die casting machines, and other plants have been avoided following their robotization. Indeed, robots have probably already saved human lives.

Identifying operational risks

Risks arising from a robotized operation are similar to those for any piece of automated, high powered, moving machinery. In a badly managed, poorly laid out and inadequately maintained set up, operators could possibly find themselves in danger of receiving electrical shocks, of colliding with moving parts, of tripping over loose laid cables, or of receiving some other unpleasant surprise as a result of the process being controlled by the robots. If a robot can be said to pose any special problems of its own, then this must be due to the nature of the control system, where a robot can start up and move without apparent warning, as a result of signals from the process interlock switching.

Let us first dispose of the electrical risks. These are no better and no worse than those for any other powered machine. Common sense dictates that the electrical power to the robot shall be supplied through switching and cabling that is installed to conform in all respects with the appropriate regulations in force locally. This must include efficient grounding of the robot body. When the robot is used to manipulate electrical tools, such as spot welding guns, then these, too, must have their cases bonded to ground. Covers protecting high voltage circuits must not be removed by unqualified personnel.

Safety rails or chains should be used to fence off active robots. Their purpose is to keep people well away from the robot's moving parts. Otherwise, visitors or other personnel might wander within range of a stationary robot, without realizing that the device was switched on, and simply awaiting a command signal that would cause it to swing swiftly in the direction of the intruder — with unpleasant consequences. As an extra precaution, emergency stops on the robot can be adjusted to mechanically restrict arm traverse.

Emergency stop buttons, fitted to the control console and (if fitted) the teach control, should be interlocked so that they stop not only the robot itself, but also the rest of the tooling in the work area. This applies especially to items such as conveyors. However, when more than one robot is working in a complete robotized system, it may be desirable to restrict the action of the stop button to removing power only in the work area immediately operated by the robot in question. The Automotive Assembly Division of the Ford Motor Company, for example, has a policy which ensures that a stop button operated in an emergency on one robot in a system, while shutting down the affected robot, does not necessarily 'power down' other robots nearby. They prefer to put a hold or freeze on the other machines, since in many cases the removal of power altogether could actually cause accidents — through memory loss, or through a robot arm sagging under the weight of its load.

Some of the Ford Motor Company's experiences in robot safety are recorded in a paper presented at the Robots 11 Conference, 1977*. This paper records the practice of restricting robot arm movement by cementing steel posts into the floor at the desired extremities of arm range. Such posts, however, are undesirable. They provide man sized pinch points, where an unsuspecting individual could find himself trapped between the arm and the post. Better to leave out the posts, when anyone putting himself in the path of the arm would stand a chance of being knocked over, but escaping without serious injury. The paper also stresses the importance of emphasizing safety practices during robot maintenance, with recommendations that all power must be locked out, all hydraulic and air power removed (accumulators relieved of pressure) and the arm blocked up on a purpose built holding device before any service work is started.

Robots with a teach/playback facility should be designed so that the robot only moves as long as pressure is applied to a trigger on the teach control — rather like the fail safe idea of a dead man's switch on a locomotive. All motion ceases when the trigger is released. During the teaching process, robots operate on a special teach mode, in which hydraulic flows are restricted to limit the maximum speed of the robot, allowing the arm to move only at slow speeds.

In some jobs, the industrial robot works with an operator who performs a complex operation on a hot workpiece which the robot holds it for him. Here, the robot would be programed to extend its arm to the maximum before presenting the workpiece to the operator. The operator would stand behind a chain or rail, facing the robot, and restricted to a location where the arm could never move closer.

Meeting OSHA (Occupational Safety and Health Act) regulations

To help insure that particular manufacturing operations are free from recognized hazards to workers, industrial robots are being used and considered for a number of jobs covered by *OSHAct* standards. Chief among these are the jobs, machines, and conditions listed under Subpart G, *Occupational Health and Environmental Controls;* Subpart I, *Personal Protective Equipment;* Subpart O, *Machinery and Machine Guarding*; and Subpart P, *Hand and Portable Tools and Other Hand-Held Equipment.*

In every industrialized country there are safety regulations with intent comparable to the USA's OSHA standards. Despite similarities, specific local requirements tend to impose design variations for robots delivered to different countries. The exporting manufacturer should clearly be aware of such legal obligations.

* *Analysis of First UTD* [Universal Transfer Device] *Installation Failures*, by Gennaro C. Macri, Ford Motor Company, Dearborn, Michigan.

Chapter 6

Organizing to support robotics

Most manufacturing enterprises are already organized to improve operations through capital investment, and productivity gains are sought continuously in well-run companies. Industrial robots are capital investments and their prime objective is productivity gain. Thus, one could presume that organizations to support robotics are in place. Why then make a special point?

Because robots conjure up images for both management and labor, and these images are likely to be false. A clear understanding of both the potential and limitations of robotics is essential to the choice of good applications by management and to the calm acceptance of their introduction by the direct labor force.

Example of manufacturer's training system

In the early 60's there were a few intuitive manufacturing executives who simply bought on a hunch. Some went on to make multiple robot installations while others failed abysmally. Today an instinctive bet is no longer a valid procedure. First of all, there are thousands of successful robot applications which have been documented by the robot manufacturers. Most of these manufacturers offer application training sessions supported by films and by hands-on experience. A typical one, based on a one-day seminar, is shown in Figure 6.1.

Armed with a few such overview training sessions, a competent manufacturing engineer can do a reasonable survey of his plant to determine the extent to which robots could be applied effectively. He can usually also enlist the aid of an experienced application engineer from one of the robot manufacturing companies.

The would-be robot user is then at a critical stage. He has learned a bit about robots and he has a feeling about their potential in his plant. No matter how fascinated he may be with robotics, if the potential applications are few and far between, he is best advised to scrub his efforts and look for other opportunities to improve productivity. Just one or two robots in a plant is probably an uneconomic investment. It's well enough to start with one, but the cost of training maintenance staff, stocking spare parts, etc. is best amortized over a group of

GENERAL APPLICATIONS SEMINAR OUTLINE

9:00 – 9:30 am   **Introduction to robots**
Welcome and introductions. UNIMATE robot background and history.
Film – 'Robots in industry'. Questions and answer period.

9:30 – 9:50 am   **The UNIMATE family**
General purpose 1000 series, 2000B and 2000C. Extended reach 2100B
and 2100C. The 4000 heavy duty unit. The R.I.G. 2000 – Welder P.P
and CP. Options and their use. Apprentice and Puma.

9:50 – 11:05 am   **Teach a robot**
Description of fundamentals of operations and functional controls,
following a pre-taught program with the program sheets and UNIMATE
robot step-by-step. Working with the UNIMATE robot demonstration
set-up.

11:05 – 11:15 am   **Stretch break**

11:15 – 12 pm   **UNIMATION'S SERVICES**
Applications engineering
    Project engineering
    Tooling design
    Systems engineering
    Training program
    Field services

12:00 – 12:30 pm   **Lunch**

Afternoon schedule

12:30 – 1:15 pm	**Foundry applications** Slides & movies Die casting Investment casting Iron foundry	**Plant tour**	**Hands on teaching time** At demonstration machine

1:15 – 2.30 pm	**Machining operations**
	System eng. & controls
	Group technology
	  (techniques)
	Parallel group set-ups
	Palletizing operations
		Plant tour	**Hands on teaching time**
			
			At demonstration
			  machine

2:30 – 3:15 pm	**Welding operations**
	Movie "UNIMATES in
	  welding" 15 min.
	Continuous path
	  capabilities
	Case histories of
	  welding operations
		Plant tour	**Hands on teaching time**
			At demonstration
			  machine

3:15 – 4:00 pm	**Talking** with the engineers about projects	**Plant tour**	**Hands on teaching time** At demonstration machine

Figure 6.1 *Outline of seminar on general applications of industrial robots,
as conducted by Unimation, Inc.*

machines. So too, if there is an obvious opportunity for numerous robot installations, top management enthusiasm can be aroused and then it will be appropriate to 'organize to support robotics' in a proper manner.

There follows a description of how one well-known manufacturing company went about this task.

How General Electric built an in-house capability

A robot induction program was developed at the General Electric Company under Mr. Vernon E. Estes of his company's Manufacturing Engineering Consulting and Applications Center. Mr. Estes is manager of Process Automation and Control Systems.

By consent of the author and his company, a detailed description of the General Electric Company's approach program is presented here* as a model of its kind.

The General Electric Company:
An organised approach to implementing robots

INTRODUCTION

The General Electric Company is a very diverse, decentralized company made up of a few large and many small manufacturing operations. Because of our decentralized mode of operation, we have corporate level consulting services organizations which are dedicated to keep up with emerging new technologies that can improve our manufacturing productivity.

Robotics technology was identified by our 'world wide monitoring' teams in 1976 as having a possible high potential for improved company productivity.

This was a relatively untapped technology in our company at this time. We knew that robots had been in the automotive industry since the early 60's, but we were now seeing an increased emphasis in applications throughout other types of manufacturing businesses, both in the United States and foreign countries.

We soon realized that this technology could put some control on one of the least controllable aspects of our machining cycles. Loading parts to a machine tool in a timely manner, at least on the surface, seemed to be the answer to many of our productivity problems. An extensive program was implemented by MEC&AC to evaluate the possible benefits of the technology from an economic viewpoint and the extent of possible applicability within the company.

This paper will cover the organized approach taken by our consulting services organization to promote robots throughout the company in an organized manner.

AN OUTLINE OF OUR ORGANIZED APPROACH FOLLOWS:

I — Evaluated economics (company wide)
II — Implemented robots in corporate MEC&AC labs (for personnel training and company wide familiarization).
Created a 'computer managed parts manufacturing' module using a robot to load and unload a computer controlled lathe.

The General Electric Company: an organized approach to implementing robots
(continued on next page)

(continued on next page)

*The following text is taken from the paper *An organized approach to implementing robots*, by Vernon E. Estes, General Electric Company. Acknowledgement is also due to consultants Block, Petrella Associates Inc., who were commissioned by General Electric to prepare a study entitled *Psychological aspects of robot implementation*. Reference is made to this study later.

III — Radiated technology
        Presentation/surveys
        Robotics implementation manual
        Seminars
IV — Psychological impact study made
V — Implementation engineers were made available throughout company
        Identify applications
        Write quote specifications
        Evaluate vendors
        Assist in planning & engineering of implementations
        Follow implementations through to completion
VI — Robot rental program established

ECONOMIC EVALUATION

In late 1976, MEC&AC sent a team out to visit various representative manufacturing operations in the company. This team was made up of two personnel, one of which was a manufacturing engineer, whose exposure to this technology was through vendor visits and a lot of reading. The second team member was an accountant. Their task was to tour shops, identify areas for possible robot applications, scope out each implementation as it would appear with a robot rather than humans and let the accountant compare the new and old modes of operations from an economic viewpoint. In the interest of saving time, only the replacement of manpower was considered as savings.

The results of this survey told us that the technology had excellent economic benefits and broad company applicability. After a presentation to our management, we were given the go ahead to start promoting the technology. Promoting in our case means more than telling people what great things this technology will do for them. The applications part of our name is important as we do not just tell our operations how to implement these new technologies, we also assist them in their implementations.

ROBOTS IMPLEMENTED IN LABS

To get ourselves educated and in a position to educate and assist in implementations throughout the company, we purchased five robots for our labs. The units selected utilized various programming techniques and ranged in price from $3,000 to $66,000.

The intent was to purchase robots which would cover the broad spectrum of available capability of those units which fit our robot definition.
Our engineers were sent to robot schools and were given time to experiment with the capability of these in-house units.

LAUNCHED RADIATION PROGRAM

In mid 1977, we launched our robotics radiation program. To be honest, we stumbled on the format quite accidentally when one of our operating components asked us to make a presentation in their plant and survey their shop to identify areas of application. This request was for a presentation which would cover 'Why robots today?', 'What are the economic benefits in G.E.?', 'What are other companies doing?', and general rules of thumb when applying. To show people that others were truly using the technology we were asked to show any films that we had covering general robot applications.

The afternoon was to be a walk through of the shop with technical representatives from the local operating component identifying simple, but lucrative (from an economic viewpoint) robot applications. At the end of the day, we were to create a list of these applications in the suggest sequence of implementation. This was a novel idea which we pursued. We reached 45 different plants with this 'one day presentation/survey' approach in a period of about twelve months.

In some plants, we identified as many as 14 possible applications that could be implemented right now.

The attendees at these presentations ranged from manufacturing engineers to general managers.

Commitments were being made on the spot by high level management to pursue the technology.

The General Electric Company: an organized approach to implementing robots
(continued)

In addition to this program, we also wrote an implementation manual which was made available within the company.

To further promote the technology, we started annual robotics seminars in 1977.

These seminars consisted of a keynote speaker, presentations by vendors and presentations by operating components throughout the company. We tried something a little unique here in that we invited General Electric people to talk about failures as well as successes. These seminars also included visits to our labs to see robots in action.

PSYCHOLOGICAL IMPACT STUDY
(This issue will be expanded upon subsequently.)

IMPLEMENTATION ENGINEERING
Our feeling is that any technology can be talked to death, but to accelerate the implementation of such a technology you may have to get your hands dirty. Our MEC&AC service operation not only tells people about the latest and greatest technologies, but also educates its own personnel to assist in the applications engineering required to make the systems work and work in an efficient manner.

The technology is unique and the success of any application of it requires knowledgeable application engineers. Our MEC&AC engineers have attended robot vendor schools, visited robot installations, assisted with implementations in the company and have had the luxury of hands-on experience in our labs.

We have found that the talk that robots cannot work as fast as humans is a myth, *if* the implementation is properly engineered.

ROBOT RENTAL PROGRAMS
To further promote this technology, we are making any robot that we have in our labs available to our operating components on a rental basis. This gives those managers who want to try something quite challenging to the state-of-the-art robot, to try before buying.

This plan is putting robots in applications that as recently as a year ago would not have been considered possible.

The General Electric Company: an organized approach to implementing robots
(conclusion)

Work force acceptance of robots

Coming now to the psychological study commissioned by General Electric, it appears that the most useful tool generated by the study for General Electric was the *Work Force Acceptance Checklist*. G.E. presumed that robotics might engender a negative labor reaction and therefore instructed their consultants to determine how to evaluate the working place environment and, incidentally, how to take action that would promote acceptance.

The advice is valid, but it will be noted that the study recognized management to be a more serious roadblock to implementing robots than is labor. Management often freezes in irrational fear of a negative labor reaction and in sheer conservative inertia. The checklist, which is reproduced in Figure 6.2, can at least alleviate the first of these issues.

Not every operation making a bid to implement robotics will be able to emulate all of what Estes has provided for his system users among the General Electric plants. Still, the fundamentals of doing application surveys, making economic evaluations, and establishing a sound working relationship with the work force should be a part of every potential robot purchaser's approach to robotics.

WORK FORCE ACCEPTANCE CHECKLIST

INTRODUCTION

This checklist is intended to enable the organization to diagnose the existing forces, both driving and restraining, which will impact *work force acceptance* of robot installation. It does *not* attempt to look at factors in management acceptance, which should be diagnosed independently, with at least as much priority. As stated in other sections of these papers, the overwhelming preponderance of interviewees indicated that management, not labor, was the roadblock. The complexity of that potential checklist is much greater and will require further study.

In the Work Force Acceptance Checklist, a maximum 'score' of one hundred is calibrated. Points may be divided between the driving forces column and the restraining forces column.

The score is determined by *subtracting* the total points allotted to the restraining forces from the total points in the driving forces column.

The result provides a barometer and moreover, enables to see if:

1. any 'new' points can be added to the driving column (while the test is discrete, management action can exceed 100 points)
2. points can be deducted from the restraining column
3. points can be 'moved' from the restraining column to the driving column by management action.

Each item on the list is assigned a specific number of points for distribution. The number of points assigned is a reflection of the estimated relative merit of each item.

An appendix is provided as an appendix to the checklist.

CHART OF RANGES

Score	Probability of acceptance
80–100	*High.* Implementation may proceed, assuming management acceptance conditions are equally high rated.
60–80	*Proceed with caution.* After examination of the feasibility of changing strength of existing forces.
40–60	*Insufficient.* Re-examination of forces and management action required to increase probability.
0 or below–40	*Failure more than likely.* A score in this range indicates a poor probability of even modifying forces.

Figure 6.2 *Work force acceptance checklist*
Developed for the General Electric Company in consultants' study of labor attitudes to robots.

WORK FORCE ACCEPTANCE CHECKLIST

ITEM	Points to be distributed	Driving	Restraining
1. Can workers be openly assured of job retention?	20		
2. Can workers displaced, but retained, be placed in equal rated jobs?	15		
3. Will the installations benefit the workers in terms of: a. health? b. safety? c. relief from dehumanizing jobs? d. relief from dirty, overly hot, back-breaking, onerous tasks?	15 *(total)*		
4. Is the present union-management climate favorable to open exchange? Disclosure of economic conditions? Labor unrest and frequent grievances? Usually distrustful? (If no union, assign points on like issues for management-work force relations.)	15 *(total)*		
5. Is the present economic condition of the organization sufficiently healthy to guarantee promises are kept?	5		
6. Have Manufacturing Engineering and other management units shown ability to establish rapport with workers or does inordinate 'social distance' exist?	5		
7. Is there management recognition and concern for the dehumanizing aspects of jobs to be performed by the robot? Or is the concern solely economic?	5		
8. Is there a plan to select and upgrade workers who will supervise or work set-up for the robot?	5		
9. Will workers on incentive rates be penalized by new rates or robot down-time not attributed to operator failure?	5		
10. Has management in the past *demonstrated* respect and regard for the talents, skills and intelligences of the workers?	3		
11. Is the organization willing to share the results of this checklist with the work force and/or union?	3		
12. Will robot training be on organizational time? Is there willingness to send the workers (if required) to the vendor's 'school'?	2		
13. Can workers express their concerns, apprehensions or fear without ridicule?	2		

TOTAL POINTS

DRIVING POINTS:

RESTRAINING POINTS:

NET SCORE:

Chapter 7

Robot economics

The success of any commercial industrial undertaking has to be measured in terms of financial performance. The most brilliant technical innovation is a failure if it results in money lost by the entrepreneur or his shareholders — or, indeed, at divisional or operating level. Robots are no exception to this rule. No matter what the social benefits are, no matter how clever the technology, no matter how pretty the robot is to watch, every proposed investment in robotics has to pass the test of a critical financial appraisal.

Investment always involves risk. Appraisal techniques seek only to reduce the uncertainties. No method exists that can remove the element of chance. A plant could become completely redundant as a result of a sudden and unexpected shift in market trends. Economic forecasts can go wrong for reasons well beyond the control of foreknowledge of factory managers. All that can be done is to use appraisal techniques with commonsense and discretion so that, all other things being equal, one proposed investment project can be shown to be more likely to prove better or worse than some other course of action or inaction.

Several techniques exist for investment appraisal. Some are extremely simple, while others are complicated to the point of absurdity. Each manager tends to adopt his own favorite method, and some of the techniques in common use are discussed later in this chapter. It is well to remember that all appraisal results rely on the accuracy of the input data, some of which has to be estimated or forecast. The most satisfactory situation is found when a relatively simple calculation shows an investment to be worthwhile by margins that exceed all possible doubts arising from shaky estimates. Fortunately, it is often found that robotics yields such positive predictions.

Whatever investment appraisal technique is used, the input data for both costs and savings must be accumulated.

Checklist of economic factors: costs and benefits

The following headings provide a framework for management analysis of the costs and benefits of the robotics installation.

1 Robot costs:
   a. Purchase price of the robot
   b. Special tooling
   c. Installation
   d. Maintenance and periodic overhaul
   e. Operating power
   f. Finance
   g. Depreciation
2 Robot savings:
   a. Labor displaced
   b. Quality improvement
   c. Increase in throughput

Commentary on checklist of costs

The following observations give further perspective to the checklist headings.

1a: Purchase price of the robot. The purchase price of a robot is highly variable, particularly if one's definition of robot includes simple pick and place devices with few articulations. The range might extend from $5,000 to $100,000 depending upon number of articulations, sphere of influence, weight handling capacity and control sophistication. Generally speaking the higher priced robots are capable of more demanding jobs and their control sophistication assures that they can be adapted to new jobs when original assignments are completed. So too, the more expensive and more sophisticated robot will ordinarily require less special tooling and lower installation cost. Some of the pick and place robots are no more than adjustable components of automation systems. One popular model, for example, rarely contributes over 20% of the total system cost.

1b: Special tooling. Special tooling can be no more than a $300 gripper for a robot installed at a die cast machine. Or, the tooling might include an indexing conveyor, weld guns, transformers, clamps and a supervisory computer for a complex of robots involved in spot welding of automobile bodies. For assembly automation the special parts presentation equipment may cost well in excess of the robot equipment cost.

Robots are not stand-alone equipment as are conventional machine tools. The interface with the workplace can be critical to success so that customers often ask that robot manufacturers bid on a turn-key basis. For economic evaluation the two prime cost factors 1a and 1b are thus combined.

1c: Installation. Installation cost is sometimes charged fully to a robot project, but it is often carried as overhead because plant layout changes were afoot anyway. At a model change there are usually installation costs to be absorbed even if equipment is to be manually operated. There is no logic to penalizing the robot installation for any more than a differential cost inherent in the robotizing process.

1d: Maintenance and periodic overhaul. To keep a robot functioning in tip-top shape, there is a need for regular maintenance, a periodic need for more sweeping overhaul and a random need to correct unscheduled downtime incidents. A rule of thumb for well-designed production equipment operated for two shifts continuously is a total annual cost of 10% of the acquisition cost. This has been borne out for thousands of Unimates many of which have enjoyed several overhauls whilst accumulating as much as 10,000 hours of field usage each.

There is variability, of course, depending upon the demands of the job and the environment. Maintenance costs in a foundry are greater than those experienced in plastic molding.

1e: Operating power. Operating power is easily computed as the product of average power drain times the hours worked. Even with increased energy costs this is not a major robot cost.

1f: Finance. In some cost justification formulas one cranks in the current cost of money. In others one uses an expected return on investment to establish economic viability.

1g: Depreciation. Robots like other equipment will exhibit a useful life and it is ordinary practice to depreciate the investment over this useful life. Since a robot tends to be general purpose equipment, there is ample evidence that an 8 to 10 year life running multi-shift is a conservative treatment.

For cost justification formulae, straightline depreciation is most commonly used, but the tax schemes of different countries may involve depreciation weighted to early years (e.g. in the U.S.A., double declining balance). Special tax credits to encourage capital investment may also influence the cost justification formula and the buying decision.

Commentary on checklist of savings

The following observations are offered on potential benefits.

2a: Labor displaced. The prime issue in justifying a robot is labor displacement. Industrials are mildly interested in shielding workers from hazardous working conditions, but the key motivator is the saving of labor cost by supplanting a human worker with a robot. So very much the better if a single robot can operate for more than one shift and thereby multiply the labor saving potential.

2b: Quality improvement. If a job is in a hazardous environment, or is physically demanding, or is simply mind-numbing, there is every good chance for quality to suffer with the mood of a human worker. A robot may very well be more consistent on the job and therefore it may produce a higher quality output. 'The robot is too dumb to mind but smart enough to do it better.'

2c: Increase in throughput. Higher quality naturally means more net output when a robot works fast enough to just match a human worker's output. However, there often are circumstances where a robot can work faster to increase gross output as well. The increased throughput is valuable in its own right, but improved utilization of capital assets may greatly supplement the economic benefit of one-for-on displacement of a worker.

Project appraisal by the payback method

Payback calculation is the simplest form of project appraisal. It depends on providing answers to the twin questions 'How much is it going to cost?' and 'How soon shall we recover the investment?'

In projects for pure research, or in ventures that are philanthropic in outlook, the initial investment may never be recovered. In normal industrial circumstances, the accountants and financial advisers will be looking most favorably on those projects which pay for themselves in a relatively short time. The time needed is known as the payback period, and this is usually measured in years. A lot depends on the type of industry, and nature of the project, but most accountants would find no problem at all in approving proposals yielding a payback period of one or two years. Beyond that, payback of three or even four years could find support.

Simple payback

An outline example of the payback method was given in Chapter 1 (Figure 1.5). This is repeated here, with slightly amended parameters.

Here is the scenario for the simple payback example. A robot is to be considered as a replacement at a workstation where 250 days are worked in a full calendar year, and where the robot would replace one human operator whose wages and fringe benefits amount to $12 per hour. Against this saving, the robot would cost $1.30 per hour to run and maintain. Capital investment, for the robot and its accessories, would be $55,000. The company normally operates one eight hour

Simple payback formula $$P = \frac{I}{L - E}$$

where

    P = the payback period, in years
    I = the total capital investment in robot and accessories
    L = annual labor costs replaced by the robot
    E = annual expense of maintaining the robot

In this example

    I = $55,000
    L is at the rate of $12.00 per hour, including fringe benefits
    E is at the rate of $1.30 per hour
    There are 250 working days per year, containing either one or two
       eight hour shifts

Case 1   Single shift operation

$$P = \frac{55,000}{12(250 \times 8) - 1.3\,(250 \times 8)}$$

     = 2.57 years

Case 2   Two shift operation

$$P = \frac{55,000}{12\,(250 \times 16) - 1.3\,(250 \times 16)}$$

     = 1.29 years

Figure 7.1 *Simple payback example*
This is the least complicated method for assessing the viability of a new project in financial terms. If the payback period indicated is very short, the result gives positive incentive for proceeding with the project, and it is not strictly necessary to use any more complicated appraisal method.

shift per day, but has the option of increasing this to two shifts per day when production demands are sufficient.

Figure 7.1 contains the calculations, and illustrates the method. If only one eight hour shift is to be operated on each of the 250 working days, it is seen that the payback period amounts to about 2.6 years. This result would probably satisfy the company's accountants. If, however, sales forecasts and production plans indicate that two shift working could be maintained, then the payback period is reduced to only 1.3 years or so, which must justify going ahead with the project with no room for doubt at all.

Production rate impact on payback

A robot does more than simply replace a man at the workstation. It might work faster or slower than the man, and it can sometimes be included in an automatic system that allows more efficient operation of

Production rate payback formula $P = \dfrac{I}{L - E \pm q(L + Z)}$

where   P = the payback period in years
               I = the total capital investment in robot and accessories
               L = the total annual labor saving
               E = annual expense of robot upkeep
               Z = annual depreciation costs of associated equipment
               q = production rate coefficient

In this example

               I = $55,000
               L is derived from a labor rate of $12.00 per hour, including fringes,
                  taken over 250 working days in a full year, each day comprising
                  either one or two eight hour shifts
               E is derived from a rate of $1.30 per hour, over the same period as L
               Z is $30,000, being 15% of the total capital cost of $200,000 paid for
                  associated machinery and equipment
               q is either 20% faster or 20% slower compared with a human operator

Case 1a   Single shift operation, where the robot is 20% slower than the
               human operator

$$P = \frac{55,000}{24,000 - 2,600 - 0.2\,(24,000 + 30,000)} = 5.19 \text{ years}$$

Case 1b   Single shift operation, where the robot is 20% faster than the human
               operator

$$P = \frac{55,000}{24,000 - 2,600 + 0.2\,(24,000 + 30,000)} = 1.71 \text{ years}$$

Case 2a   Double shift operation, where the robot is 20% slower than the human
               operator

$$P = \frac{55,000}{48,000 - 5,200 - 0.2\,(48,000 + 30,000)} = 2.02 \text{ years}$$

Case 2b   Double shift operation, where the robot is 20% faster than the human
               operator

$$P = \frac{55,000}{48,000 - 5,200 + 0.2(48,000 + 30,000)} = 0.94 \text{ years}$$

Figure 7.2 *Complex payback example*
These calculations should be compared with those for the same project that were
carried out with the simple payback formula, in Figure 7.1. In this particular
example, the inclusion of capital utilization and production factors for the work
station has had a marked effect on the results.

one, two or even more pieces of expensive machinery. These realities
are taken into account when the production rate project appraisal
method is used. These are the value of capital equipment in the application and the production rate coefficient compared with the manual
worker standard.

In this more complex example, assume that the total capital value of associated machinery and equipment is $200,000, and that the company takes an annual write down percentage of 15% depreciation in the operating cost budget. This gives an annual depreciation of $30,000, represented as Z in the complex formula. Production rate variations are covered by a production rate coefficient, q, which is the rate by which robotized production either exceeds or lags that achieved by a human operator. Here, a rate of 20% above human rate, and a rate 20% below human rate are to be tested in the calculation, giving values for q of plus and minus 0.2, respectively.

The actual calculations are given in Figure 7.2. It is interesting to compare the results with those obtained from the simple payback calculation, and to see how the consideration of production rates and associated capital utilization can reduce or extend the expected payback period. Assuming that the robot is reliable, and has no inherent bugs, the shortened payback result should apply to most robot applications, where the robot can operate continuously without rest periods, and with dependable repeatability. Thus, for one shift operation, the 2.6 years indicated by the simple study becomes only 1.7 years in the complex example. Similarly, the two shift result is reduced from 1.3 years to a payback period of only 0.9 years (all results to one decimal place, since the accuracy of estimates does not justify more significant figures).

Return on investment evaluation

Payback analysis works best when the overall timescales under consideration are short. Changes in money values owing to inflation, or notional or real rates of interest applicable to project financing are generally ignored. Special purpose automation, custom built to produce one workpiece, is prone to early obsolescence. Provided that the payback period is suitably short compared with the expected equipment life, such factors as inflation or annual interest rates are unlikely to weigh heavily on the argument for giving the project a yes vote or a thumbs down. This is very obviously true when payback periods of only one or two years emerge from the calculations.

Robots are not special purpose automation. Their flexibility means that they can be redeployed when a product line changes. Their reliability indicates a long working life — at least eight years in the case of Unimate robots. It is not, therefore, unreasonable to regard robots as general purpose equipment. Since the investment is going to produce useful work over a period of many years, changes in the value of money, interest payable, and the rate of return on the money invested all provide factors that can be evaluated and considered in a project go or no-go decision. The simplest way to approach this problem is to decide the rate of interest or other return that company policy dictates for its investments, and then ensure that the proposed new project at least comes up to these expectations.

For the return on investment (ROI) calculation illustrated in this chapter, the proposed project is the same robot acquisition as that used in the payback examples. All figures and operating conditions are the same, but one additional factor has been introduced. The total value of the investment in robot and accessories is to be written off by equal instalments over the life expectancy of the robot — eight years. The ROI calculations are demonstrated in the columns of Figure 7.3, and

Parameters for return on investment example

Investment in robot and accessories	I =	$55,000	Number of working days per year	=	250
Annual depreciation, eight years, straight line	=	$ 6,875	Robot replaces one human operator		
Hourly cost of robot upkeep	=	$ 1.3	Return on investment (ROI) =	$\dfrac{S \times 100}{I}$	per cent
Annual savings resulting from robot	=	$ S			

	8 hours per day operation		16 hours per day operation	
A Robot costs	$		$	
Annual depreciation	6,875		6,875	
Annual upkeep	2,600		5,200	
Total annual robot costs	9,475		12,075	
B Corresponding labor costs				
Wages, including fringes, at:				
$ 8 per hour	16,000		32,000	
10 per hour	20,000		40,000	
12 per hour	24,000		48,000	
15 per hour	30,000		60,000	
C Annual cost savings and ROI				
$ 8 per hour	6,525	ROI = 11.9%	19,925	ROI = 36.2%
10 per hour	10,525	19.1%	27,925	50.6%
12 per hour	14,525	26.4%	35,925	65.3%
15 per hour	20,525	37.3%	47,925	87.1%

Figure 7.3 *Example of return on investment calculation*

the results are shown graphically in Figure 7.4. By any standards, the predicted return rates are impressive. Even single shift working looks good. A big contributory element in these encouraging predictions is the longevity and freedom from early obsolescence demonstrated by robots, without which the initial investment could never have been written off in such small instalments, over such a long period.

Color plates

1. Combined robotic and visual inspection system from Auto-Place, Inc. in a testing and inspection operation for auto industry. Installation raised productivity 400%.

2. Standard Auto-Place Series 50 robot on a double slide loads and unloads parts from a track on to a chucking machine.

3. Electrolux MHU-Senior robot engaged in heat treatment for auto industry.

4. Electrolux MHU-Senior robot serving injection-molding machine, with MHU modules for secondary operations.

5. In an aircraft manufacturing application, the Cincinnati Milacron computer-controlled T^3 industrial robot presents wing panels to a machine for drilling and riveting.

6. Two Cincinnati Milacron T^3 robots work together handling refrigerator liners. Liners are made of thin-gauge plastic and can be extremely difficult to maneuver. First, one robot removes a liner from a moving conveyor and places it in a trim press. Then, at the other end of the trim press, another robot removes the liner and hangs it on another moving conveyor overhead. Both robots use Milacron's tracking option to follow the conveyors, maintaining continuous knowledge of where the refrigerator liners and hooks are located, and thus maintaining the correct relationship between the robot arm and each refrigerator liner and conveyor position.

7. ASEA robots cutting ingots at Kohlswa Steelworks, Sweden.

8. ASEA robots spot welding at Saab-Scania, Sweden.

Figure 1

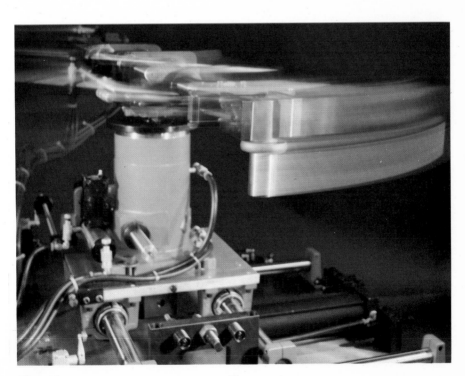

Figure 2

Figure 3

Figure 4   iv

Figure 5

Figure 6

Figure 7

Figure 8

Figure 9

Figure 10

Figure 11

Figure 12

Figure 13

Figure 14

Figure 15

9. Unimate handling hot metal billet in foundry operation. In the Birmingham (UK) plant of TI Tubes Ltd. this heavy duty robot is feeding an Ajax upsetter from an 18-billet rotary hearth furnace. With one man to oversee it the robot handles in a shift 400 forgings for automotive axle cases and hub-ends. When the billet is the right temperature of 1100°C the robot places it in a descaler, then into the Ajax (in the foreground) which it triggers sequentially through up to four forging operations. After forging the robot places the billet on to a conveyor and repeats the cycle.

10. Unimate engaged in die casting at Bruce Manufacturing, Conn.

11. Unimates in action: auto spot welding.

12. Continuous path welding by Unimate at AiResearch.

13. Stamping operation by Unimate in auto frame manufacture, Ford Motor Co.

14. Unimate handling glass at General Electric plant, Jackson, Miss.

15. The Unimate line making turbine blades by investment casting at Pratt & Whitney.

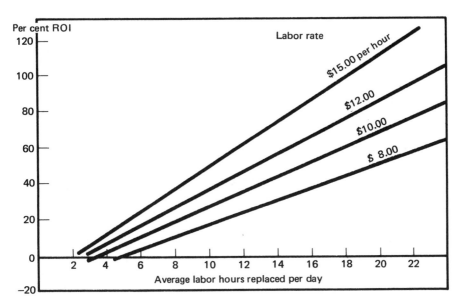

Figure 7.4 *Return on investment graph*

Areas of cost exposure

OK — so where's the catch? Isn't there a flaw somewhere in this good news? In fact, although there is no real catch, two counter points must be declared, either of which can depress the optimism shown by the payback or the return on investment calculations.

The first word of caution concerns the level of robot upkeep costs. Figures used in this chapter are applicable to a company which already employs some robots, and which has therefore developed the capability and organization for maintaining and reprogramming robots. If the new robot is going to be a lone newcomer, the first robot in a total population of one, then the company is going to have to set up a maintenance facility from scratch. All costs of that maintenance operation are going to be lumped against the single robot, adding both to the initial investment required and the annual upkeep figures. If the new robot is the second to be acquired, then the situation looks a lot rosier, with the initial investment virtually nil for the maintenance facility (since it already exists), and with the total upkeep costs for both robots amounting to something far less than double the costs for the existing single robot. It has to be said that, for practicable and economic purposes, robots are a far better proposition when the robot workforce numbers upwards of three or four.

The second counter argument against favorable results calculated so far in this chapter only becomes significant when expenditure or returns are considered over timescales measured in two, three or more years. For very short payback periods, there is no hidden snag, and no corrections have to be applied. When payback periods are long, or when return on investment predictions are based on writing the investment off over several years, then discount factors have to be taken into account. Discounting recognizes that fact that a dollar spent today is more expensive than a dollar spent next year, since by delaying the expenditure the money can be kept in a credit account accumulating interest. Conversely, money earned next year is worth less than the same sum of money earned this year. These effects obviously increase with the number of years involved.

The application of the discounted cash flow concept to investment appraisal of robot systems is excluded from this book. The reader is referred to the textbooks of managerial finance, where the principle is treated in detail.

Chapter 8

Sociological impact of robots

The following anecdotes are drawn from real life and dramatize the opportunities which robots offer to relieve humans of degrading working conditions.

O In a Bridgeport, Connecticut factory, a market survey team was led into an assembly press department. Most of the workers were women in middle age or older. Each person sat at a work station where part *A* was assembled to part *B* in order to produce part *C*. This operation was noisy. There were no less than 150 work stations in the shop, so that the racket was indescribably painful — not even Muzak could penetrate the din.

The foreman outlined his ideas of what a robot would have to do in order to enter his work force. Then, he took one of the surveyors on to one side and shouted confidentially into his ear:

'See those two women over there? When they come to the end of every shift, I have to take them by their shoulders and tell them it's time to go home. If I didn't, they'd go right on working into the next shift. We've robotized them all right! All I've got to do is get the parts out, but I wish we could do these stupid jobs without people.'

O The president of a small die casting business candidly discussed his employment problems:

'My father hired German immigrants. They took great pride in coaxing a cantankerous machine into producing good zinc die castings. By the time we added automatic temperature control and automatic ladling, the second generation workers would have no part of tending a die casting machine.

'So, we turned to the underprivileged negro for our labor force. Now, the only people we can get to face the physical abuse of die casting are newly arrived Puerto Ricans. Sooner or later, they will opt out too, and only robots will be able to stand the gaff.'

Quality of working life

No one will dispute that robots offer unique advantages to those workers who must otherwise spend all of their working day in conditions that include noise, vibration, smells, smoke, excessive heat or intense cold, oil spray, flying chips, monotony, or risk of serious personal injury. To continue with the die casting example, a robot replacement can soon be taught to operate the die casting machine. The human worker can then be transferred to another job in the plant, where he no longer has to breathe die lubricant fumes, where hot molten zinc no longer spurts at him, where asbestos gloves are not the standard uniform, and where he can work as a man and not as an eight hour per day *de facto* robot.

Social improvements usually conflict with lack of available funds. Robotics is often the exception to this rule, so that moving workers into jobs with better conditions is associated with increased productivity at the robotized work stations. Die casting and press operation are just two of the many examples where moving machine parts pose real physical danger to those who have to tend the machines. There must be many other jobs, not only in the manufacturing industries, where people are at risk, and where robots could take over their duties. It can be argued that those working to produce and market industrial robots have a moral duty to give priority to those specific areas in industry where men and women are being degraded by the nature of their work, or put into actual physical danger.

As development continues, and robots become more and more adept, a large segment of humanity will be released for endeavors more worthy of human beings. Edwin Markham might have written his melancholic poem *The Man with the Hoe** about a man at a punch press. Should anyone ask in the years ahead:

'Whose was the hand that slanted back this brow?

Whose breath blew out the light within this brain?'

The robot claim — both financial and humanitarian — should be pressed until every debilitating job is delegated to one of these sub-human slaves.

Attitudes to robots

People, in general, are suspicious of change. Whether a plan is for a new road, a new airport, or the introduction of automated plant in a factory, someone, somewhere is going to object. Even a new employee is going to be regarded askance, until he has proved that he can fit in with the rest of the team. Such opposition is the result of fear, and fear is born of ignorance. Every worker wants to be assured that change will not threaten his working conditions adversely, reduce his earnings, or even make his job redundant. If management fails to inform its workers of forthcoming changes, keeps people unaware of the reasons for those changes, and does not discuss the effects of proposed changes on individual livelihoods, then of course people become afraid, resentful,

*Edwin Markham, *The Man with the Hoe and Other Poems*, New York (Doubleday), 1912.

upset, worried and hostile.

Robots are a special case in this respect. They bear some resemblance to human beings, because they can act autonomously and simulate some human actions. Through fiction, men have long been taught to treat robots with a wary eye. When a robot is introduced into a factory, there would appear to be every reason for a mass walk out. The three anecdotes from actual experience which follow contradict this view.

Case 1: Appreciation of robot effort

In a Cleveland, Ohio sheet metal stamping plant a robot struggled with an unwieldy automobile dash board. Although it tried hard, the robot just could not make the production rate demanded by the foreman. When they tried to increase the working pace, the robot hand dripped oil and dropped parts. The regular, human workforce looked on. They were fascinated by the Herculean efforts of the robot, and the dogged attention that it received from its harried programmer.

A few circuit changes improved the robot morale, and a new layout simplified the job. One morning, two months later, the robot performed as required, the production rate was achieved, and then exceeded. It settled down to a shift in, shift out work routine, and when it finally made the rate it earned a standing round of applause from every man in the department.

Case 2: Get-well cards for sick robot

There is an automotive plant in Chicago where a robot joined a brawny, hard working blue collar crew. This machine had something akin to a nervous breakdown, and it pulled in its arm, thereafter refusing to move. Routine maintenance by plant personnel evoked no response from the robot at all. A specialist was called in. By the time this specialist had diagnosed the illness, prescribed the treatment, and effected a cure, a full week had elapsed.

This was long enough to galvanize fellow workers. 'Clyde the Claw' was a colleague in distress. The men of Production Department No 14 organized a get-well party, and heaped cards and flowers at Clyde's pedestal. In fact, the whole gang posed for a photograph, arms draped over the robot's dormant frame. This picture, and the story, were carried in the company newspaper under the headline *Sick Clyde*.

Case 3: Blue-collar support for robot productivity

A major die casting company today uses 37 industrial robots to tend die casting machines in one of its plants. The company faced a crisis when there were only five robots in this plant. Union leadership demanded the right for union labor to do the programming. At the outset, management resistance was vociferous; programming a robot surely required engineering training. Management certainly could not trust a member of the bargaining unit to program for peak production.

The roboticist called in to advise in the ensuing arbitration found it difficult to defend the management position. 'Even our company's directors can teach our robots. Who then can't?' was his somewhat irreverent argument.

To the chagrin of the plant manager, the issue was settled in the union's favor. But, to his amazed relief, the union operators took great pride in coaxing the utmost out of every robot on the line. Odd, how men who would pit all of their innate ingenuity against machines at which they were harnessed could turn completely around and devote those same talents to maximizing the output of a slave worker.

Acceptance out of affection

The case histories show qualities that never occurred to market survey analysts in the early days of robotics. The currently miniscule robot population has already interacted with human co-workers and the union leadership. Initial curiosity was supplanted by tolerance, and finally there has developed something very much akin to rapport.

As the evidence mounts of man's remarkably benign reaction to robot workers, it is interesting to speculate on the underlying causes. An industrial robot is pathetically sub-human. It is deaf, dumb and blind. The emotion most clearly aroused in human colleagues is one of amused affection. The robot is obviously a latter day slave and, better still, it is a willing slave. The self-evident inferiority of a minority group has often been the ethical justification of slavery. Master races have been deeply embarrassed by the intellectual prowess of their slaves, when they begin inconsiderately to display all the attributes of a peer group. A robot slave could never be guilty of such an affront. It offers no challenge.

The roboticist is painfully aware of the magnificence of even the lowliest specimen of *homo sapiens*. There is no hope for creating a slave robot in man's own image. Fortunately, there is also no need. What does a factory worker need with the senses of taste, hearing and smell? Indeed, in human workers these senses often become atrophied in self defense against the hostile factory environment. In the unpredictable world where man has evolved, the robot is but a hapless, obedient, musclebound oaf. These are qualities that evoke pity rather than resentment. Robots are therefore truly likeable.

Evaluation of worker attitudes

The reader is referred back to the Work Force Acceptance Checklist set out in Chapter 6 (Figure 6.2). It was recognized by the consultancy study which generated the checklist that management conservation or hesitancy were more serious blocks to a robot installation than worker rejection.

Effect on employment

The chapter on Robot Economics clearly proposes that the driving force behind adoption of robots in manufacturing plants is cost saving. While quality improvement and increased throughput produce some of the savings, labor displacement is the central benefit. Robots contribute to productivity primarily by displacing human workers. The benefits are clear; what about any sociological costs?

Even if no individual can point to a particular robot and say that this machine has taken his job, it is surely true that jobs have been eliminated by robots. The impact is miniscule to date. By the end of 1979 there were some 6000 robots installed worldwide. On average, considering that many robots work multi-shift, the displacement may total 9000. The net employment impact of the formation of a robotics industry may actually be positive in creating jobs — if one considers the numbers working to develop and produce robots and the numbers working as suppliers to this new industry.

Thus, robotics has no history worthy of extrapolation to project a sociological cost of job displacement by robots. But, all the technology and specifically all of automation over the past century can be evaluated on a cost/benefit basis. And one will find in favour of technology despite the emotional protestations of the hair shirt back-to-nature advocates. While philosophy is not the province of this book, the position is taken that gains in productivity are always good.

Consider productivity gains in U.S. farming. In 1870 47% of the U.S. population was engaged in food production; by 1970, 4% was all that was required to produce all of the food needed for the nation as well as to create a vast surplus. There were dislocations during that 100 years as people left farms for the cities. Still, the benefits have greatly outweighed the costs.

Back to robotics. What can be reasonably expected for the industry for the balance of this decade? Many independent surveys seem to be arriving at an average annual growth rate of 35%. A fair estimate of the robot population increase for 1980 is 2000 machines. At a 35% p.a. growth rate the population will be increasing by some 40,000 per year by 1990. The current technology will not support such a growth rate, but it is a conservative expectation that by 1985 the robots will boast of at least rudimentary vision (see Future Capabilities in the next chapter).

Forty thousand robots entering the work force per year! that's a big number, but not a disruptive one. It represents about 0.06% of the blue collar work force in the industrialized countries where these machines would be employed. Of course, there will be other automation influences on productivity, as well. These will probably eclipse the impact of robotics since automation is already so much a part of the manufacturing scene. And, automation that is unwaveringly addressed to specific segments of manufacturing is more likely than robotics to have a disruptive 'pocket of displacement' influence on employment.

Robots will enter the work force gently. There is no job activity concentration. Witness the spectrum of suitable jobs in Part II. Productivity that can be achieved without massive displacement is the most socially acceptable. Even though all productivity *is* good, a distributed influence that impacts no faster than natural attrition can be most readily accommodated socially.

Labor leaders in the USA will ordinarily support productivity improvement so long as it does not come from some disguised form of speed up or does not result in massive displacement. One other caveat is that the union membership shares in the economic benefit. This is not an unreasonable stance and robotics is compatible therewith.

In Germany much is made of 'humanization of the work place'. Jobs are to be enriched and made safer. Robotics fits into this pattern because many jobs are outside the pale of enrichment. Even in the Comecon countries robotics enjoys intense interest. In principle in those socialist states there is no unemployment. There is only the question of gross output from the total population, however employed. The elimination of debilitating jobs becomes a social cause; unemployment is not at issue.

In short, robotics will contribute importantly to the material well-being of mankind, without painful dislocation of individual workers. If 50 years from now the work week is three days, the air and water are clean again and the industrial life is ever so desirable, we shall be at least partially beholden to robotics.

Chapter 9

Future capabilities

Whatever the intentions of their creators, robots are always going to be compared with men in terms of their attributes and general behavior. Although they might be just fine for doing repetitive, dirty, boring, dangerous jobs in factories, and even though they can often do such jobs with positive economic advantages, yet robots remain stupid, insensitive and limited devices when they are compared with human beings.

No robot can hope to match man with his acute senses, ability for free thought and judgement, artistic appreciation, capability for self reproduction, efficient conversion of food into energy and body cells, and properties of recovery from many illnesses and injuries. The gulf between man and robot will always remain, but, although it cannot be closed, this gap is going to be reduced as technology advances. This chapter examines some of the characteristics that might be developed in the foreseeable future and goes on to discuss how these additional or improved attributes might affect the general application of robots.

Future attributes of robots

Figure 9.1 is a recapitulation of the principal attributes to be found

1   Work space command with six infinitely controllable articulations between the robot base and its hand extremity
2   Teach and playback facilities — realizing fast, instinctive programming
3   Local and library memories of any practical size desired
4   Random program selection possible by external stimuli
5   Positioning accuracy repeatable to within 0.3 mm
6   Weight handling capability up to 150 kilos
7   Point-to-point control and continuous path control, possibly intermixed in one robot
8   Synchronization with moving workpieces
9   Interface allowing compatibility with a computer
10 Palletizing and depalletizing capability
11 High reliability — with not less than 400 hours MTBF
12 All the capabilities available for a price which allows purchase and operation within the traditionally accepted rules for economic justification of any new equipment

Figure 9.1 *Robot qualities already commercially available*

among the range of successful robots in manufacturers' current catalogs. The last item listed is by no means the least important. Any manager who has a possible robot application is obviously going to satisfy himself not only that the potential robot recruit to his workforce is capable of doing the job, but also that it can be bought or hired for a price that stands up to scrutiny by at least one of the recognized techniques of economic evaluation.

Features which are considered to be very desirable goals for future robots are listed in Figure 9.2. State-of-the-art technology already places several of these characteristics within the grasp of the robot designers, and some are to be seen on experimental robots operating in laboratories. The more sophisticated these devices become the more they will obviously cost. But, trends towards continuing or even increasing labor cost inflation favor such developments, so it is likely that such advances will become economically viable sooner rather than later.

---

1   Rudimentary sense of vision to provide
     a) recognition data
     b) orientation data

2   Tactile sensing giving
     a) recognition data
     b) orientation data
     c) physical interaction data

3   Computer interpretation of the visual and tactile data

4   Multiple appendage hand-to-hand coordination

5   Computer directed appendage trajectories

6   Mobility

7   Minimized spatial intrusion

8   Energy conserving musculature

9   General purpose hands

10 Man-robot voice communication

11 Total self diagnostic fault tracing

12 Inherent safety (Asimov's Laws of Robotics)

13 All the capabilities above available for a price which allows purchase and operation within the traditionally accepted rules for economic justification of any new equipment

---

Figure 9.2 *Robot qualities sought for the future*
This list of features is either possible or rapidly becoming possible with state-of-the-art technology. When they have been developed to the point where they are both reliable and economic, then they should provide considerable extension to the scope of robot applications in industry and beyond.

If the features listed in Figure 9.2 are difficult to achieve, some of them become even more elusive when attempts are made to combine them with other features of the same robot. For example, it is possible to mount a small video camera at the end of a robot arm to provide some sort of visual signals, but this would work against Item 7 on the list, minimized spatial intrusion. Another important factor in any new development is the effect on robot reliability. Unless special care is taken in the design and selection of components, and in the quality of construction, added complexity will down-grade the statistical probability of failure from the 400 hours MTBF already achieved.

The problems notwithstanding, it is reasonable to expect that the attributes of Figure 9.2 will be attained. They will be attained because the technological demands are not outrageous and the merit of the need is clear. One can evaluate prospects for success from the viewpoint taken by the United States Department of Defense in a 1967 program titled 'Project Hindsight'. The conclusion was that innovations happen when there is a juxtaposition of:

1 A recognized need
2 Competent people with relevant technology
3 Financial support

Given all the attributes which robots already offer and the bank of experience that existing robots have accumulated, there is a mounting pressure for additional capability. This is what the Air Force study called 'a recognized need'. Worldwide, competent people are joining the fray. Robotics is a great fun game and there is immense opportunity for satisfaction in making a contribution. The people who elect to become roboticists bring with them a relevant technology.

Kinematics	Structural engineering
Dynamics	Tribology
Servo design	Metallurgy
Fluid power	Metrology
Digital electronics	Sensory instrumentation
Analog electronics	Character recognition
Computer structure	Industrial engineering
Integrated circuit design	Manufacturing engineering
Computer software	Physiology
Cybernetics	Bionics
Automation technology	Psychology
Numerical control	Sociology
System engineering	Economics
Rotating machinery	Futuristics
Gear design	Oceanography

Figure 9.3 *Disciplines useful to the robotics game*

The relevant technology is broad indeed. Figure 9.3 catalogs some of the relevant disciplines. The more successful roboticists may very well be technological generalists. The industrial world has the competent people with the relevant technology; the need is recognized; and the third ingredient, financial support, is coming from a myriad of sources. First of all, there are the robot manufacturers who devote a percentage of their revenues to advanced research and development. These pioneers have been joined by government organizations who sponsor research and development in the public interest. Gains in productivity and the release of man from onerous tasks are considered to be in the public interest. Grants go to universities, non-profit research laboratories and even to industrial concerns. The Comecon countries likewise divert a portion of their wherewithal to robotics research and development. The amount of activity generates increasing pressure and the attributes listed in Figure 9.2 become urgent needs for those with finance, and attainable needs for those with the technical expertise. Most, if not all, of these attributes will be on hand before the end of the 1980s decade.

Commentary on future attributes

In the listing of robot qualities sought for the future (Figure 9.2) there is reason to add some explanation. The explanations may reflect upon the technological demands of the problem or they may reflect upon just how urgent is the need for this particular attribute. Right off, it must be noted that number 1 on the list *is* number one. Rudimentary vision will have the most profound effect in broadening the application base of robotics. Right behind vision is tactile sensing and this, too, is a grand enhancement of capability. None of the others are capricious suggestions. They too will have their impact.

1: Rudimentary vision

Vision will allow a robot to recognize things and also to determine where things are. These are listed as recognition data and orientation data. In the factory environment, orientation data is much more important. Most factory managers know what it is they are dealing with. There are times these parts may be mixed in an industrial environment, but by and large the manufacturer can tell a robot what parts it is looking at. The problem remains, time and again, to determine *where* these parts are.

A human idiot can look into a basket of parts and pick them out one at a time and orient them for placement in a secondary operation. This idiot work has so far eluded the artificial intelligence community. In academia this is called the 'occlusion problem' or the 'bin picking problem'. If one has a bin full of identical parts, how does one direct the robot to pick them out one at a time in oriented fashion? This problem has enjoyed the most intense research and development efforts and this intensity is entirely justified.

Pattern recognition researchers have classically devoted their attentions to recognition and that technology has become extremely useful in medical instrumentation and in the business paperwork world. But in the factory, recognition is not the key because production control departments will be able to tell people or robots what parts are where and when. A wag at Ford Motor Company expresses it this way: 'I never expect to see a Chevrolet come down a Ford line.'

At the moment, experimental vision systems are pretty good at defining the position of an individual black part on a white background. That is a good first step because it is sometimes possible to isolate individual parts that are disoriented. However, that is the exception. The factory world is not black and white, it tends to be brown and gray and it tends to have subtle gradations of these dull colors.

Solutions will come through making robot vision more sophisticated, making the algorithms for analyzing the camera scenes more elegant and by rationalizing the workplace. The last is most important. If a manufacturer recognizes that robot vision is, indeed, rudimentary, he might be able to help the robot by maintaining some level of order in the transport of parts throughout a factory. Palletized parts are much easier for a robot to find than are parts dumped random-scramble in a box.

Research is feverish, and with every justification, because researchers have grand vision instrumentation to play with and they have the privilege of deploying digital electronics with casual largesse since this art has not yet reached its cost plateau. Cameras are cheaper, resolution is higher, computers are more powerful, faster and cheaper, and the humans who write the algorithms are getting smarter.

When a robot can look into a box full of parts and extract and orient them one-by-one, it will have fully arrived. Meanwhile, a sighted robot can be useful at picking up separated parts on a conveyor or at finding holes in a structure, in which holes are not precisely located.

Specialized vision applications will be practically applied early in the 1980s and sophisticated systems that can interpret 16 levels of gray in a scene will be commonplace by the end of the decade.

2: Tactile sensing

After dwelling so long on vision, it is important to give proper deference to tactile sensing. Indeed, if this technology moves faster than the vision technology, it could become more important early on. It is well established that a blind person can be most effective in a number of activities that depend upon tactile sense. So, too, for a blind robot with tactile sensing capability. A robot with tactile sense could use its capabilities to recognize parts just as most humans can detect what something is in a darkened room. By groping about the human is reasonably capable of determining the orientation of things.

But, for a robot, the more important tactile sense is the sense that tells the robot somehow what is going on during the interaction of parts or tools in its hand with the workpiece before it. This has been listed in Figure 9.2 as physical interaction data. We all know the experience of putting a nut on a bolt in a blind location. We almost instinctively back-thread until we feel a click and then we run the nut forward, ever sensitive to the possibility of encountering jamming if we should get a crossed thread. This is a tactile sensing experience we would wish to add to the sensory perception repertoire of a robot.

Both vision and tactile sensing are not absolutes. There will be gradations of sophistication and the exquisite scene analysis capability of the human being is unlikely to be matched. Indeed, it is unlikely that such development will even be sponsored because emulating the full sensory capability of the human being is not likely to be considered a recognized need and, therefore, it is unlikely to gain financial support.

3: Computer interpretation of the visual and tactile data

Perhaps this item should not be isolated from Items 1 and 2. But it is evident that a videcom camera scene is readily interpreted by a human being. If the robot still has difficulty, it is because of the inability of the robot's control system to analyze this extremely clear scene before it. Scenes are digitized, thresholds are formed, the data is fed to the computer. Algorithms for interpreting this vision scene are critical. Those who ultimately write programs that enable a vision module to give precise location data to a robot will have made a most profound contribution.

4: Multiple appendage hand-to-hand coordination

The cowboy who fills his cigarette paper and rolls it with one hand is much to be admired, but there are many industrial jobs that are really much better handled by a two-handed operator. So, one of the needs for robotics is choreography that will enable two robot arms to work in consort just as do the two arms of human assembly workers.

5: Computer directed appendage trajectories

As the finishing touches are being put to this book, it is embarrassing to note that computer directed appendage trajectories are in fact now available for commercial robots. Rather than delete this item, we will let it stand as a symbol of the volatility that is to be found in robotics research and development. Point-to-point programming was available at the very outset of robotics, but doing such things as straight line insertions and extractions have been most tedious because robot articulations are not geometrically likely to permit the desired action with the motion of a single articulation. Trajectories in any direction in space require compound collective motions of a robot's arm.

6: Mobility — the roving robot

For the bulk of factory robot jobs, the robot stands in a single station just as does his human predecessor. In fact, the factory job that has an operator moving busily outside a two-meter radius is probably inefficiently designed. Still, there are jobs for roving operators who must tend stations that are widely separated and only need service periodically. For this, mobility is needed and to date this mobility has been delivered in a heavy-handed fashion. That is, robots have been mounted on rails to travel between work stations. What is needed is a robot that can literally stroll.

7: Minimized spatial intrusion

Most robots require substantially more floor space than do their human counterparts. This can actually eliminate the potential use of the robot because the cost of laying out equipment in a new fashion and providing for factory floor space can kill the economic justification for using the robot in the first place. We designed factories for many years to match the human physique. A challenge today must be to produce a robot that matches the human physique because this may literally be easier than to produce manufacturing plants matched to the spatial demands of current-day robots.

8: Energy conserving musculature

There is an elegant research and development or invention statement to be extended to the robotic community. Robots today use substantially more energy to accomplish their work than does the incumbent human operator. Rarely does the human operator deliver more than 1/20th h.p. to accomplish his job and it is not uncommon for a robot to soak up 3 h.p. to 10 h.p. in the same activity. The ratios speak out tremendously in favor of further technological contributions. Energy use is not yet an abiding issue in robotics, but it can become so and the comparison between robot-power drain and human-power drain is a compelling argument for improvement.

9: General purpose hands

Human hands are simply awesome in their capability. Thus far, no robot activity has attempted to emulate the human hand. Rather, special-purpose hands have been developed and built to match specific tasks. Robot manufacturers with many machines in the field have an extensive library of hand designs upon which they can draw to meet customer requirements.

Usually, if the job changes, the hand must be changed. This is not an incapacitating handicap. When a robot is reassigned, there are probably many other application changes going on and a hand redesign can be accomplished in a timely fashion. However, there are jobs where a single hand is just not appropriate to the entire robot task. The 'heavy handed' solution (no pun intended) is simply to give the robot a selection of hands and a quick disconnect stump wrist. The robot then selects from a nest the appropriate hand of the moment. This is rather akin to a numerically controlled machine selecting from a bank of cutting heads.

However, if time is of the essence, it is distressing to have a robot continually return to a home position in order to exchange hands. One would prefer to have a more universal gripper which could cope with a range of gripping activities. It might have three opposing fingers and these might each have articulations of their own. Tactile sensing might broaden the flexibility of such a gripper. Invention may be necessary and the call surely will not go unheeded.

10: Man-robot voice communication

To date man-robot voice communication has been something of a gag, since programming robots is not all that arduous and the robot's intelligence has not yet advanced to the point where a few subtle suggestions will motivate the robot to take desired action.

On the other hand, a growing library of robot programs is being amassed. The robots are gaining adaptive features such as rudimentary sensory perception. The robots will in due course be granted a model of their surrounding environment. With so much in hand, it may be attractive to allow the human boss to use plain English in instructing a robot as to its ongoing work. Moreover, the robot which is likely to be highly sophisticated could, with justification, respond to the human boss with synthesized speech to explain its view of the work situation. Or, its speech might be used to explain internal ailments which need service attention.

Fortunately, the technologies involved in speech recognition and in speech synthesis are growing in sophistication and decreasing in cost. They will be useful tools for man-robot communication.

11: Total self-diagnostic fault tracing

The trend toward sophistication in robotics is inexorable. And, thus, as the robot becomes easier and easier to assign to factory tasks, it will become ever more internally sophisticated. Whatever the level of robot sophistication, it is crucial that the machine exhibit an on-the-job reliability competitive to that of a human worker. Thus, the robot user must have a long Mean Time Between Failure and a short Mean Time To Repair. If the machine is, indeed, an elegant one, then repair will be intellectually demanding. What is needed and what will be provided is a self-diagnostic software package that pinpoints a deficiency under any failure condition and directs the human service staff in efficient methods for recuperating performance.

12: Inherent safety (Asimov's Laws of Robotics)

Asimov's Laws have been already recited. They become more important as robots become more competent and as robots are utilized in more intimate relationship with other human workers. Safety must be inherent if robots and humans work shoulder-to-shoulder with the robots doing the drudgery and with the humans contributing the judgement. The development task is not easy, but fortunately it is also not impossible.

13: All the capabilities above for a price which allows purchase, etc.

Once again, as was mentioned in the listing of existing robot qualities, robots must offer these additional capabilities and still be economically justified. If a system is devised to give a robot a dazzling visual capability, but if this system is a multi-million dollar research accomplishment, it has no place in industry. The robot plus all its accessory attributes must be just as easily economically justified as is any other prosaic piece of cost saving equipment. This contraint is often missed completely by academics, but the marketplace remains intractable. No economic benefit — no market.

Priorities in attribute development

Having listed all of the desired attributes, it is important to repeat again that these will, by and large, be all made available during the decade of the 1980s. It is even more important to emphasize again that of the missing robotic attributes, the two crucial ones are vision and tactile sensing.

We all know that human vision serves its possessors in a spectrum that ranges from the near-blind to 20/20 vision. Twenty-twenty vision accompanied by 20/20 ability to analyze scenes is not in the cards for robots in this century, if ever. But, each advance that permits a robot to peel away a cloudy curtain and thereby understand its surrounding environment will enhance the robot's utility.

It is a simple vision system indeed that tells a robot whether or not an opaque item is present or absent. Photocell devices to accomplish this binary task have been available for decades. A more advanced vision system can detect not only presence or absence, it can also identify an object that is present. The next step is to not only indentify, but to determine the position and orientation of an isolated part, perhaps a black part on a white background. Then, the robot's eye, by increasing discrimination, can detect a part that is one level of gray against the background of another level of gray. Finally, the robot may be given the ability to discriminate among a number of gray parts that are in juxtaposition and perhaps even obscuring one another. At each stage of sophistication new opportunities will arise for robots in the factory.

The importance of being able to see and to interpret what is seen and to react intelligently to what is seen cannot be overemphasized. The workplace is being rationalized. More and more often, factories will take pains to preserve orientation, but when the robot can cope with disorientation, then the application potential will burgeon. This eyesight evolution will move ever faster during the decade of the 1980s and every triumph will be accompanied by a geometric progression in robot utility.

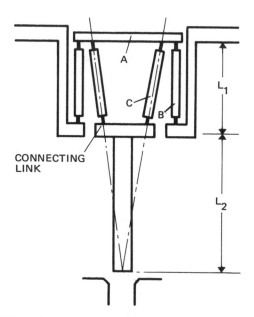

Figure 9.4 *Compliance device for mating parts*
The vertical links (B) give a compliance center effectively at infinity. So lateral forces produce only lateral motion. The other links (C) give a compliance center at the end of the peg which allow only angular rotation in response to a moment applied at the tip.

The second most important frontier is tactile sensing and here invention has already occurred. Draper Labs in its efforts to computer control the interaction between parts which must be mated came upon a serendipitous conclusion that such parts-mating can be eased by a completely mechanical passive accommodation. The device known as the Remote Center Compliance (RCC), is shown schematically in Figure 9.4. It is already being used experimentally by researchers attacking the problem of programmable assembly.

Sensory perception is getting full measure of attention by industrialists and academics. Consider the typical laboratory facility depicted diagrammatically in Figure 9.5. This is the facility being utilized by SRI International in its exploration. The system has the hierarchy of computer capability, it uses vision input and wrist force sensor input and voice input. Throughout the industrial world similar facilities under the command of bright technicians are being used to create the missing algorithms of sensory perception.

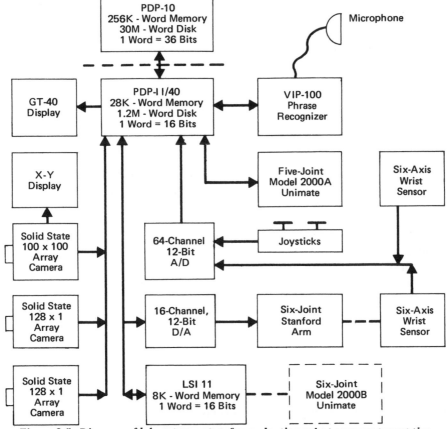

Figure 9.5 *Diagram of laboratory setup for evaluating robot sensory perception and manipulator dynamics*

Interaction with other technologies

It so happens that factory managers today who may be considering robotics are also confronted with a rich array of other new manufacturing technologies that individually promise impressive gains in productivity. The opportunities are almost overwhelming in their diversity. Moreover, the benefits are more often theoretical than proven and conservative manufacturers choke up when confronted with hard investment projections weighed against soft productivity projections.

Adoption of all of the advanced technologies listed in Figure 9.6 is probably inexorable, but if the history of numerical control and more recently of robotics is a measure, then we may expect plodding, grudging acceptance. Note that numerical control is firmly entrenched. Though it is the 'establishment' technology, its adoption is a very far cry from saturation.

○ COMPUTER AIDED DESIGN – CAD
○ COMPUTER AIDED MANUFACTURE – CAM, NC, CNC, DNC
    Factory Data Management
○ GROUP TECHNOLOGY
○ AUTOMATIC INSPECTION
○ AUTOMATIC WAREHOUSING
○ AUTOMATIC ASSEMBLY
○ ROBOTICS

Figure 9.6 *Advanced technologies contributing to productivity improvement*

Taken altogether, the technologies listed in Figure 9.6 make real the prospect for the long heralded 'unmanned factory'.

If the industrial robot could literally stand in for factory blue collar workers on a one-for-one basis, then robotics would be no more of a driving force for advanced technology than is the current employee complement. As it is, robots are subhuman indeed. They dearly need the kind of factory environment promised by all the other blossoming technologies. Robots, to become ubiquitous, need rationalization of the factory.

Now, if any of the other advanced technologies can be justified in its own right, the economics become that much more compelling should acceptance happen to grease the way for robotics.

Group technology

Consider group technology as a beachhead. If parts are classified into families and machine tools are then aggregated into complementary families, the handling of parts throughout the manufacturing process becomes robot work. Parts en route to finished goods are never dropped into tubs for interdepartmental transfer or buffer storage. Their orientation can be maintained and that is a touchstone of machine tool loading and unloading by robots.

Robots already can load and unload machine tools, change tools, palletize and depalletize, but only people can locate and orient parts stored random-scramble in tubs.

Group technology, with or without robots, cannot really stand alone. Full benefit requires the factory data management subset of Computer Aided Manufacture (CAM). Manual production control systems cannot cope with product flow through a plant laid out for group technology. The production control system needs computerization for rapid response. Harassed production control clerks protect themselves with buffer storage and that demands reversion to those tubs and the frustrating orientation problem. Even ignoring the demands of robot part handling, the great blessing of group technology is the reduction of in-process inventory. If inventory is not computer controlled, this primary benefit is lost.

Automatic warehousing

Probably next behind NC in successful installations is automatic warehousing. Computer-controlled order picking is almost conventional in systems that boast great volumetric efficiency. Automatic warehousing can be justified by the combination of real estate saving and labor saving, but available systems are not quite compatible with robotics.

In most automatic warehouses, material is tracked in and out as to general location (that is, a standard pallet); but for a robotized factory not only location but orientation should be preserved. With an order-picking robot riding the warehouse trolleys, parts could be stored and reclaimed at the detail level and then delivered to work stations in known orientation. At the moment, all order-picking for major automotive subassemblies is done manually.

Automatic inspection

Given automatic warehousing and a tight factory data management system, automatic inspection begins to tie in with robotics. A robot with test equipment akin to that ordinarily in use by human operators could inspect at its work station and then segregate in accordance with inspection data. The factory data management system would 'escort' inspection data through the entire in-process cycle. Thus, parts arriving at robot work stations would have signatures as to key qualities and orientation. Knowing where a part is and really knowing what it is enables a blind robot to cope, be it in forming, machining or assembly operations.

These are isolated examples of 'escort memory' already being used in robotized lines, such as automatic steel casting production and auto body spot welding, but to the author's knowledge there is no pervasive data management system escorting total production flow. Robots would like that.

Automatic assembly

Programmable automatic assembly is being attacked in a number of laboratories. By and large, the engineering disciplines being invoked are those of the roboticist. Variable mission automatic assembly, as distinguished from classical hard automation, may boil down to robotics at its most elegant. There is great concentration of labor in medium-run assembly which is not susceptible to hard automation. But this kind of operation should succumb to robots having just a bit more sensitivity than is available in today's commercially available industrial robots.

CAD-CAM

This leaves computer-aided design and computer-aided manufacture to be tied into the robot bailiwick. They have generally been lumped together as CAD-CAM and they are a forlorn duo being primped up and touted, but never finding a friendly bosom. It may be recalled that a robotic argument has already been made for computer-aided manufacture or at least that part concerned with rationalizing data flow through the factory. Robots also opt for NC, CNC and DNC. If the parts are being formed under computer control, their entrance into and exit from the process are clearly work for robots. There is no human artistry being aborted by robot handling.

Computer-aided design can be powerfully defended. A CAD data bank protects a manufacturer from reinventing the wheel, and it speeds up the design process — witness suspension bridges and circuit boards. A CAD-robotics filter would spit out all the no-nos of robot handling or robot assembly to slap designers' wrists when they design parts that could embarrass robot workers.

Stand-alone versus *distributed robotics*

During the 1970s everyone had the chance to enjoy the movie *2001: A Space Odyssey*. The closing passages are delicious in allowing each of us to plead privileged insight as to the deeper meaning. But earlier on there is less opportunity for wild-eyed speculation. We find humans in mortal combat with a robot. His name is Hal, but we never see him. He pervades the space ship. The humans attack his distributed intelligence and his communication channels. Like an octopus, he does not die with the loss of one tentacle.

Is this the inexorable direction for robotics? Or could our future robot be a pottering artisan, self-contained machine doing pretty much as his human forbears had done? Certainly in the spectacularly successful movie *Star Wars,* the two stand-alone robots, R2D2 and C3P0, struck an empathetic chord in American imagination that was never accorded to HAL.

One solution is to change the very nature of the workplace through computer aided design and computer aided manufacture, CAD/CAM, which is really numerical control reaching back to the drawing board, and reaching forward to cutting tools and factory management. At the same time, the workplace would become highly stylized through group technology classification of parts and siting of equipment. Orientation is preserved. The manufacturing system becomes so highly rationalized that the peculiar adaptability of the human being is no longer necessary. Or, necessary only for such menial tasks as 'sweeping cuttings'. Such is the report coming from the McDonnell-Douglas parts fabrication plant in St. Louis. Note where the rationalization went out of the job — in the jumble of cuttings on the floor.

CONCEPT OF THE UNMANNED FACTORY

The unmanned manufacturing plant is a virtual practicality in machined-parts manufacturing. From East Germany through the USA to Japan there are isolated showplaces where batch manufacturing is carried on automatically under computer control. Rarely, if ever, has one of these systems been economically justified at the time of installation; but, the advance expense will inevitably serve these pioneers well in the future.

The situation is not so rosy when it comes to operations other than machining. Foundry work, forge shops, and the most labor-intensive activity of all, assembly, are not succumbing easily to automation except in very high volume manufacturing.

Enter the robot! But what kind of robot? Will it be an extension of DNC machining center technology with hardly any anthropomorphic associations, a HAL, or will it be a lovable stand-alone robot, a C3P0, using appendages, sensory perception and resident intelligence to carry out man's behest in the factory?

Probably it will be a combination of the two. Even without robots, the computer is leaning heavily on the production process. Production control is rapidly being computerized and telecomputing feeds back the status throughout the factory floor. Inventory is managed, machine loading is decreed and the workers get their daily job assignments from the computer.

HIERARCHY OF CONTROL SYSTEMS

Let's continue the argument with reference to Figure 9.7 showing a

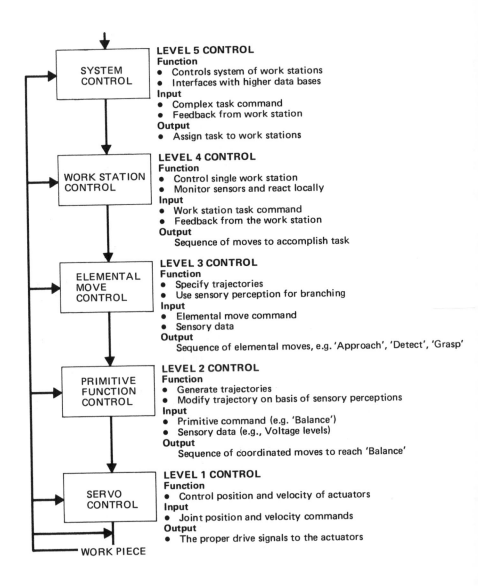

SYSTEM CONTROL

LEVEL 5 CONTROL
Function
• Controls system of work stations
• Interfaces with higher data bases
Input
• Complex task command
• Feedback from work station
Output
• Assign task to work stations

WORK STATION CONTROL

LEVEL 4 CONTROL
Function
• Control single work station
• Monitor sensors and react locally
Input
• Work station task command
• Feedback from the work station
Output
  Sequence of moves to accomplish task

ELEMENTAL MOVE CONTROL

LEVEL 3 CONTROL
Function
• Specify trajectories
• Use sensory perception for branching
Input
• Elemental move command
• Sensory data
Output
  Sequence of elemental moves, e.g. 'Approach', 'Detect', 'Grasp'

PRIMITIVE FUNCTION CONTROL

LEVEL 2 CONTROL
Function
• Generate trajectories
• Modify trajectory on basis of sensory perceptions
Input
• Primitive command (e.g. 'Balance')
• Sensory data (e.g., Voltage levels)
Output
  Sequence of coordinated moves to reach 'Balance'

SERVO CONTROL

LEVEL 1 CONTROL
Function
• Control position and velocity of actuators
Input
• Joint position and velocity commands
Output
• The proper drive signals to the actuators

WORK PIECE

Figure 9.7 *Hierarchical control system for robot installation*

hierarchical control system. This concept originated in the National Bureau of Standards and was adopted by the U.S. Air Force as an integral part of its ICAM (Integrated Computer Aided Manufacturing) philosophy. Five levels of the hierarchy of control are shown. One presumes that the hierarchy continues on up to the cosmic computer in the sky where the really big decisions are made.

This hierarchical system prevails in its counterpart, the DNC machining center, and the extrapolation to robotics is rational but perhaps a bit too seductive.

There will be operations linked to a computerized data base that will permit precise definition, both of the product handling and the manufacturing context. For the time being, these manufacturing situations will be very much in the minority. Even where it is all logical on paper, it will take uncharacteristic boldness on the part of a mangement to make the investment commitment. Without unrestrained government support, it might not happen at all.

But let's go back to Figure 9.7. If we could build a stand-alone robot to replace humans in a large percentage of their more puerile factory roles, how high up the hierarchy must we go? From the definitions, level four should be high enough. Of the fourth level, NBS says that 'this level of control takes care of the complete operation of a robot in its associated work station'. When we hire a human operator, we ask no more. The heritage of industrial engineering as espoused by Ford and glorified by Taylor breaks the work down into simple tasks, easily learned, and imbued with no skill-level bargaining clout. We don't give human operators job responsibilities as broad as the fifth-level. This is part of the industrial engineering-supervisory role, a white collar job. It is enough in the stand-alone robot business to strive for a broadly useful blue collar worker. He needs no more than level-four intelligence and manual dexterity.

Figure 9.2 has listed the attributes the stand-alone robot still needs to function well at level four.

It should be noted that getting a robot to level-four in a subhuman sense is where the action is. All of the higher levels are being developed anyway for manufacturing systems that have people at level-four, and on down.

Future applications

En route to the full level-four robot as conjectured by the National Bureau of Standards hierarchy, each addition of attributes, or enhancement of attributes, will open up new application opportunities. It is safe to conjecture regarding some of these that are on the threshold of being realized. However, it may very well be that the existence of an advanced robot with enhanced attributes will occasion new applications that as yet have not aroused the speculation or roboticists. There is historical precedence since many current applications were hardly in mind during the robot development phase. One thinks, for example, of investment casting which has become an application exceptionally suited to robots with currently available attributes.

One field that is sure to be carved out by robotics is assembly. This is the one future application which will be considered in detail since it is imminent and because it enjoys such intensive development effort today. Before discussing assembly, however, it might be well to describe briefly some prospective applications which have been suggested and for which there is more than a glimmer of hope for success.

1: Cleaning parts. The requirement is to remove randomly located flash from plastic parts as well as metal parts. The inelegant solution of cleaning an entire surface whether it needs it or not is too time-consuming, but the robot with sensory perception would be able to pick out the areas in need of attention.

2: Automotive paint spraying with absolutely no human presence. If every human being is eliminated from the automotive paint spray booth, then conditions for spraying paint can be optimized and the problems of health hazard and environmental protection can be eliminated. This application is being vigorously attacked by a General Motors team of engineers who have developed an extremely sophisticated paint spraying system that involves robot arms. One could conclude that GM is leaning to the distributive vs. the stand-alone robot.

3: All kinds of packaging and, specifically, packaging that requires vision. An example under current study is the packaging of chocolate candies which arrive at a packaging station in disoriented fashion and which must be found, oriented and nested in candy boxes at a high rate of speed.

4: Electrical harness manufacture. Traditionally, electrical harnesses are made on a 'harness board' and human operators lead wires around pins to specified destinations after which these wires are bundled to complete the harness. Every harness has its own board and this may involve vast amounts of tooling storage. Work is under way to automate this process with a robot arm to lead and bundle the wires. Programmable automation will be used to create a universal harness board.

5: Package distribution. Loading of trucks to distribute packages is a harsh task which may succumb to robotics in conjunction with a supervisory computer to explain to the robot how the packages should best loaded to achieve a high packing density.

6: Handling soft goods. The robot with both visual and tactile sensing and perhaps a universal gripper may be able to help the hard-pressed garment and shoe industries which are very labor-intensive and whose work load is slowly being relegated to the Third World.

7: Sheep shearing. Sheep shearing is seriously being considered for robotics by an Australian concern that has devised means for immobilizing sheep during the shearing process. The sheep-shearing robot must have contour following capability and it must have force sensing capability. As the investigators say, if this application were successful, robots would have finally entered a 'primary industry'.

8: Prosthesis. Work has been done to build extra-skeletal structures around humans who have lost control of their limbs. This has not been very successful to date and, at best, it is a travesty of human dignity. Another solution is to put a robot under voice command of the paraplegic, thereby giving the paraplegic the full-time benefit of an automaton servant. This 'Man Friday' concept would provide a physical extension of the unfortunate handicapped person without the need for the one-on-one emotional strain of a physically complete human continually serving the handicapped individual. It is possible, it is worthy, and it just might obtain sufficient financial support to become a reality.

9: Service industries. An officer of MacDonald's once asked Unimation if a robot could produce the hamburgers and the french fries and the Eggs McMuffin. Unimation engineers demurred, but with a bit more rationalization in the fast-food business, it might be possible to put sensate robots to work and then hire the youngsters just to entertain the clientele.

Another service application might be the collection of garbage. With garbage delivered curbside in standardized containers, a garbage truck could roll along the street under human command while a robot at the tailgate would pick up, empty and return garbage containers to the curbside.

10: Household robot. Even a household robot may be practicable before the end of the 1980s decade. Given the advanced attributes and a house which is designed to match the needs of both robot and human inhabitants, we might bring the servant class back into being.

Applying robots to assembly work

Closer to reality than all the foregoing, however, is the use of robots in assembly. In the discussion of Robotics *versus* Hard Automation, it has been noted that a great deal of assembly is already done with special-purpose automation. On the other hand, there is a vast amount of assembly work which does not lend itself to special-purpose automation. The bulk of assembly is still done by people and some 40% of the so-called blue collar workforce is engaged in assembly.

Professor Boothroyd of the University of Massachusetts asks why assembly work still requires human attention. He concludes that the bulk of batch production assembly operations will not enjoy conventional automation for a variety of reasons. He points out that conventional automation generally involves a special purpose one-off machine and therefore cannot be considered for assembly of products other than those satisfying all of the following requirements.

○ A volume of at least one million per year.
○ A steady volume of production.
○ A market life of at least three years.
○ A size of the order of between 0.5 and 20 inches with individual parts to be automatically assembled generally between 0.05 and 5 inches in their maximum dimensions.
○ Consisting of parts which do not deform significantly under their own weight or will not break when dropped from a height of about 3 inches onto a hard surface.
○ Parts quality must not require human operator adaptability.
○ Part on part, 'pancake', assembly must be possible.

These are clearly serious constraints and since we do get all of our various assemblies together, it must be concluded that humans are not so constrained. The hope now is that robots with their new attributes will also not be constrained and therefore able to take over batch production assembly operations.

There are many excellent investigators addressing themselves to the problem. In addition to Professor Boothroyd at the University of Massachusetts, there are groups at Draper Labs; SRI International; Rhode Island University; The National Bureau of Standards; The Institute for Production and Automation in Stuttgart, Germany; Purdue University; Westinghouse (under National Science Foundation sponsorship in its Automatic Programmable Assembly System (APAS) development), and many more. This is not an exhaustive list and other bright investigators are joining the fray.

A program of more than routine dedication is the General Motors program that carries the acronym PUMA, Programmable Universal Machine for Assembly. In this effort, Unimation Inc. is a General Motors subcontractor. General Motors specified a robot with space intrusion comparable to that of the human being and with a weight handling capacity of 5 pounds. Surprisingly, the GM analysis indicated

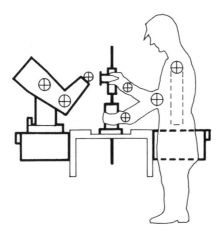

Figure 9.8 *Robot designed to human size*

that 90% of the parts used in an automobile weigh less than 5 pounds. Figure 9.8 is the GM spec drawing for a robot designed to human size. Figure 9.9 is another General Motors sketch indicating how these robot arms might be introduced into an otherwise quite conventional indexing assembly line.

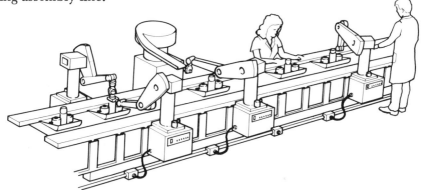

Figure 9.9 *Introduction of robot arms into conventional indexing assembly line*

Figure 9.10 is a portrait of the Unimate 500 robot which has been designed to meet the requirements of the General Motors PUMA system. It is expected that PUMA systems will be able to do automotive subassemblies such as dashboard, taillights, window cranks, transmissions, speedometers, carburetors, alternators, etc.

Of importance to the ultimate success of this venture will be the VAL language which is embedded in the Unimate 500 robot's

Figure 9.10 *Human size robot – the UNIMATE 500*

computer control. The functional attributes of this very powerful language are listed in Figure 9.11. It should be noted that this language equips the robot to accept and act upon sensory perception signals be they visual or tactile. Originally, the GM program expects to use robots without sensory perception, but as this equipment becomes available, the PUMA system should be able to cope with ever-more sophisticated tasks.

1	Operates in real time
2	Programmed in joint, world or tool coordinates
3	Contains editing to add, delete or replace
4	Built-in diagnostics
5	Speed scaling
6	Programming interaction while operating
7	Instructions for looping, branching, position indexing
8	Interfacing outputs and inputs
9	Automatic approach and departure
10	Program compensation for workplace orientation
11	Off-line floppy disk program storage
12	Continuous path operating at constant tool speed
13	Line tracking capability
14	Sensory perception interface-closed loop and iterative override
15	RS232C computer interface

Figure 9.11 *Attributes of VAL computer language for addressing PUMA robot*

There is no end to the exciting prospects for enhancing robot capabilities and thereafter broadening the robot's application achievements. Thus, a book entitled *Robotics in Practice* will need regular supplementation to remain up-to-date and useful. One hopes this chapter hints at just how much fun lies ahead for later editions.

PART 2
Application studies

Industrial robots have already demonstrated capability in a broad range of tasks. The heart of this book must be discourse on the proven roles for robots in the factory.

For each robot application, the operation or process is first described in detail and then the robot is introduced. If a work station must be modified to accommodate the robot, such interface requirements are outlined, as are any peculiar robot qualities that are demanded by the job.

Die casting was the first process to be robotized. This first case will enjoy the most exhaustive treatment and to a certain extent serves as a generic application chapter. Subsequent application descriptions may deal more briefly with robot characteristics that have already been described in one or another phase of the die casting application. Beyond this, each application is complete in itself.

Chapter 10

Die casting applications

In the die casting process, parts are formed by forcing molten non-ferrous metals under pressure into metal molds called dies. Alloys of lead, aluminum, zinc, magnesium, copper and brass are commonly used. The die casting machine consists mainly of two heavy platens, one fixed and one moving, which accommodate the dies, these normally being fabricated in two halves. The whole design is massive enough to withstand the very high pressures used, typically thousands of pounds per square inch.

Outline of die casting operation

In operation, the die halves are closed and locked automatically under pressure generated either from air cylinders or by hydraulic means. Molten metal is then delivered to a pump which may be cold or heated to the temperature of the molten metal according to the method used. The plunger of the pump is advanced to drive metal quickly through the feeding system while air in the dies escapes through vents provided for this purpose. Sufficient metal is introduced in each such shot to overflow the dies cavities, fill any overflow wells built into the dies and to produce some surplus metal or 'flash'. The pressure is maintained long enough to allow the metal to solidify, after which the die opens to permit the casting to be ejected.

Ejection is usually by means of pins, built into the dies, which extend as the die opens, forcing the casting out of the machine. Sometimes manual operation is necessary to free a casting completely. It is also essential to keep the dies well lubricated to prevent the casting from adhering to them and to provide a better finish. Also a correctly chosen lubricant will allow metal to flow into cavities in the die which might otherwise not be filled. Lubrication is usually carried out by an operator spraying the die surfaces between shots although the procedure can be automated. Cleaning of the dies is also a necessary routine, especially to remove unwanted scraps of metal which might prevent the dies from closing on the next shot with resulting damage. Air jets are frequently used for this purpose.

After removal from the machine, the casting is often quenched by transferring it to a bath of water. In the simplest case the casting simply falls from the machine into a water bath on ejection. From the quench tank, the casting is placed in a trim-press which removes all excess metal.

Types of die casting machine

Two types of die casting machine are in common use, hot-chamber and cold-chamber designs. If the metal being cast melts at a low enough temperature so as not to attack the injection pump materials, then the pump mechanism can be placed directly in the molten metal bath. This is the hot-chamber machine. However if the molten metal is prone to attack the pump materials at the casting temperature, then the only resort is to place the pump outside the metal bath giving rise to the so-called cold-chamber machine. In this type of machine molten metal must be transferred or ladled between the bath and the injection system, usually by manual means although the process can be auto-mated. Cold-chamber machines are usually confined to the use of aluminum, magnesium or brass.

The principal components of both types of machine are illustrated in Figures 10.1 and 10.2. In either type, the shot of molten metal may weigh anything from a fraction of an ounce up to about 50 pounds according to the size of the casting. In hot-chamber machines rates of up to 500 shots per hour can be achieved, but the rate with cold-chamber machines tends to be slower because of the extra time needed to ladle metal between the bath and pump system. Another limiting factor which increases cycle time and lowers the production rate is the set-time, that is the time required for the shot of molten metal to cool

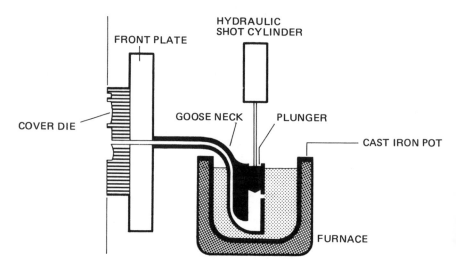

Figure 10.1 *Elements of the hot-chamber die casting machine*

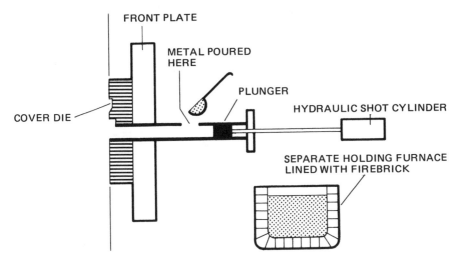

Figure 10.2 *Elements of the cold-chamber die casting machine*

sufficiently for the casting to be ejected. This is obviously dependent on the size of casting, and where necessary, water cooling of the dies is used to reduce this delay.

Factors affecting the casting cycle

Die maintenance of all forms, including lubrication and cleaning is an essential but time-consuming operation. Any form of cleaning by a human operator may require him to place his hands inside the dies with the risk of a serious accident should closing of the dies occur through error or malfunction. In addition, the die casting environment is not only risky, but unpleasant due to the presence of molten metal under pressure, often in a cramped and hot work place.

Despite the need for certain manual operations, on average some 75% of the cycle time of hot-chamber machines is automatic. With cold-chamber machines however only 45% of the cycle is controlled, the rest being taken up with manual operations. Numerous factors influence the cycle time of die casting machines such as variations in the quality of metal used (causing changes in operating temperatures) machine and die wear, frictional problems, etc. Whenever human operators are involved in any part of the cycle, further variables are introduced due to the operator's personal work-habits which are generally unpredictable. However an experienced operator will always achieve significantly better results from his machine than anyone less conversant with the installation and process.

The molten metal flows into the dies through a series of channels cut into the die surfaces. The main entry port is called the sprue, and when the casting cools, the sprue, which is fully charged with metal on each shot, forms a stub which is useful as a place to be grasped when the casting is removed from the machine or quench tank and placed in the trim-press. When casting in aluminum, the sprue is referred to as the biscuit because of its disc-like shape.

From the sprue the metal passes through one or more narrow channels called gates, and from these to other channels or runners which link up the entry port with all parts of the die to ensure that it is fully charged with molten metal on each shot. When the casting cools, these runners remain as metal rods or bars which may conveniently be used to grasp the casting at all stages until they are removed in the trim press. To increase production, whenever possible more than one part is cast in the same die. In this case the individual sections of the multiple mold are joined by runners which are eventually trimmed off. Surplus metal is collected from the trim press for recycling through the metal bath. Figure 10.3 illustrates sprues, gates, and runners in a simple but typical casting.

It is standard practice in a die casting shop for the sequence followed to be (1) casting, (2) removal, (3) quench and (4) trim. All of these are 'pick and place' operations which have a bearing on the use of robots in this application.

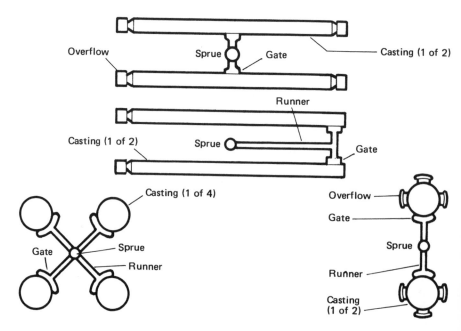

Figure 10.3 *Die casting operation: sprues, gates and runners*

Robots in die casting

The die casting industry has been a pioneer in the use of robots, the first being introduced as early as 1961. Since then more than two million hours of operation have been accumulated in shops throughout the USA and abroad. In many instances these machines work 7 days a week, 24 hours a day on long and short runs. One venerable robot – a Unimate – has already exceeded 90,000 hours of die casting production. That amounts to 45 manshift years in a tedious, hot and dirty job justifiably spurned by man!

More and more die casters are using robots to keep up with modern foundry practice and competition. Unimate robots at work in die casting far outnumber all other robots combined, and they work in both custom and captive shops, round the clock and single shift operations, manipulating light and heavy loads.

While robots have been used in a variety of modes in die casting they have without exception provided high net yields, better and more consistent quality, reduced die wear and lower costs. Downtime due to malfunctions average less than 3%. On short-run jobs where changes in the robot's program are required frequently, die-setters have found that they can teach the robot new routines after only a short period of instruction. In the more aggressive shops, up to 12 die casting machines can be serviced by six robots, all under the supervision of one operator.

Both hot and cold-chamber machines can be served by robots. Although the robot can be programmed to ladle metal in the cold-chamber machine, several factors favor the use of other techniques, most of them automatic. However the use of robots in the industry has highlighted the need for the development of more durable and reliable ladling systems to allow the robot to realize its full potential. The early success and subsequent widespread use of robots in die casting was for sound reasons. The application offered several features favorable to robots. First, consistent and precise part orientation is essential in using robots, and this is inherent in die casting. Second, minimal changes in existing equipment are needed when a robot is installed to take over from human operators. Typically a robot can be operating two days after delivery to the shop. Next, quite simple and usually standardized fingers can be used to grasp the castings by the gate or sprue. The ability to change jobs with a minimum of teaching time has already been mentioned. Here are some of the jobs which robots are now successfully handling in die casting shops.

1: Unloading one machine, quenching the part and disposing it

This is the simplest and least sophisticated usage, and as a result it probably offers the lowest economic return. However, it should not be ignored. Short casting-cycles are required for small, lightweight parts cast in hot-chamber machines, and they provide the robot (and for that matter the human operator) little time to do anything other than dispose of the part. However, production rates have been increased by 20% by virtue of the more consistent cycling of the machine arising from the robot's constant rate of working.

Yields of usable castings can be increased by some 15% simply because the regular rate of operation results in more constant and uniform die temperature. The result is much less scrap. In this application, the robot is shown in Figure 10.4 in its installation which allows it to unload the casting machine, quench the casting and deliver the part to a conveyor. Cycle time is controlled by the die casting machine. The robot is interlocked to await the completion of a shot and the opening of the moving platen before reaching in to grasp the sprue. The interlocking circuitry is shown in Figure 10.5. It is this system which keeps the robot in synchronism with the die casting machine. Since the production rate will be slow compared with the speed of the robot, there will be a considerable dwell between each shot during which the robot is inoperative. This is not economical.

As a safety precaution the casting can be monitored as it leaves the

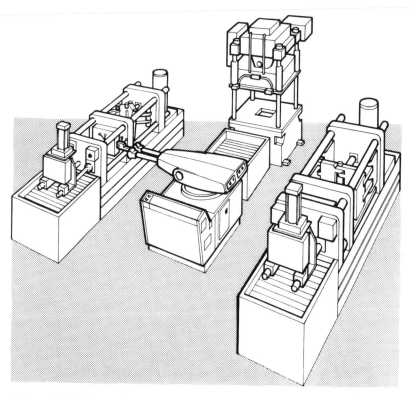

Figure 10.4 *Die casting installation to unload, quench and dispose of part*

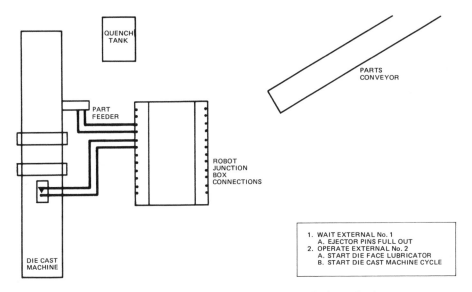

Figure 10.5 *Circuitry for unload, quench and disposal of part*

die casting machine to ensure that it has been completely removed and no broken pieces are left in the dies. This can be accomplished by having some extremity of the casting brush against an external limit switch as the casting moves towards the quench stage. The switch can also be arranged to shut down the operation and signal for maintenance attention if some part of the casting is left in the dies.

More sophisticated monitoring devices using infra-red detectors are now available which can provide a high degree of certainty that the casting has come cleanly out of the die.

2: Unloading one machine and rough trimming

This operation is similar to the one just described except that a trimming operation is included, usually to separate the rough casting into several different parts. Rough trimming is required when multi-cavity dies are used to cast several parts simultaneously, after which they must be separated, sorted and finally trimmed and finished. Quenching is often unnecessary, and careful positioning in the separating tool is seldom required, so high production rates can be maintained. In addition there are no pieces of trim and flash to dispose of since the operation merely separates the parts from the sprue and the runner.

Commonly the robot maintains its grip on the sprue while it is returning to pick up the next casting. The application can be refined somewhat by the provision of a simple conveyor on which the sprue and runners can be dropped and returned to the furnace for recycling.

3: Alternate unloading of two die casting machines

The long reach capability of robots like the Unimate makes them adaptable to unloading two die casting machines alternately. This mode of operation is generally employed where larger castings and longer machine cycle times are involved. Depending on the length of these cycle times it may be possible to include trimming or separating in the robot's sequence. However this is not often attempted for good reasons. First, either two separate trim presses would be required or identical dies would have to be installed in both die casting machines at all times. Second, if a single trim press were used the robot might be unable, within the limitation of the usual five articulations, to orient the casting properly in the trim die because it would be unloading from the right and left alternately. (A sixth axis would be required.) Third (and this is true whenever a robot is used in trimming) downtime of the trim press must result in downtime of the whole die casting station. In view of the relatively low capital investment of a trim press and its ability to work at much higher speeds than the die casting machine or the robot, it is considered by most users to be economically unwise to jeopardize maximum output from the die casting machine by tying its operation to the trim press.

For unloading two die casting machines in parallel an important characteristic of the robot is its ability to have more than one program at its command. For example, the robot may be taught to unload the die casting machine on its left then unload the machine on its right, and to alternate between them. When there are die changes required on either machine, the robot should be able to be directed to operate the single operational machine. The set-up man is in full control. He can quickly shift the operation from one machine to the other or to an alternating sequence between the two. The rationale applies if either machine is stopped for any reason such as malfunction or repairs. In addition, with appropriate sensing devices and interlocks, all of a simple nature, the robot can be taught to respond automatically to a fault signal from one machine whereupon it will continue to service the other machine only. Should a fault occur in the second machine before double service has been restored, the robot can be programmed to shut down all operation.

This mode of operation achieves maximum utilization of the robot, particularly in the independent shop where short runs may place a premium on set-up time. Once the robot has been taught a basic program for each of the two die casting machines, a die change requires, at most, a slight adjustment of the pick-up point to cater for the changed sprue-location provided by the new die.

4: Unloading one die casting machine, quenching the part and trimming it

A dynamically superior robot can unload a die casting machine and quench the part at a rate of about 500 shots per hour. When larger castings are being produced, the production rate is much lower than this, so the robot has plenty of idle time which can be put to productive use. Thus if a trim press is moved within reach of the robot, the secondary operation of trimming the casting can be accomplished well within the cycle time of the die casting machine.

Long run jobs are especially suited to this operation since then initial set-up time is a minor factor both in time and cost. The sequence has particular attractions when the part requires no further work after trimming, such as plating, inspection or hand work, for then the part can be packaged for delivery to the customer.

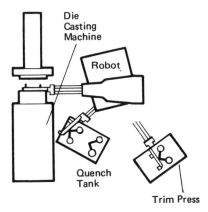

Figure 10.6 *Equipment layout for die cast unload, quench and trim*

Figure 10.6 shows the layout of a robot at a die cast and trim station. In such a layout the prime synchronizing signal is generated on shot completion and this originates from the die casting machine. Command signals are typically provided not only for purposes of synchronization but also for use in safety circuits.

A switch senses whether or not a complete casting is in the robot's fingers. If not, the operation is stopped. Figure 10.7 is a circuit diagram which is used to tie together the die casting machine, trim press and robot.

Automatic trimming poses a number of problems, the most serious being scrap removal. Removal is usually accomplished by strategically located air jets which blow away flash which cannot be disposed of by gravity. Scrap disposal can be rather difficult with more intricate

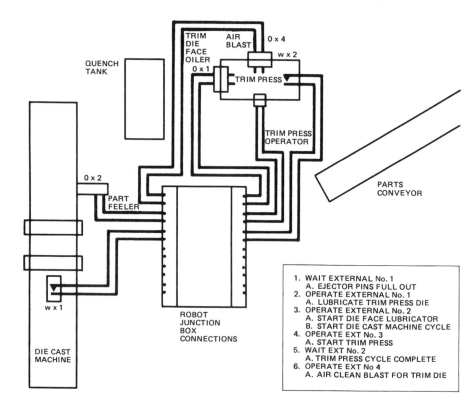

Figure 10.7 *Circuitry for unload, quench and trim*

casting forms. The effectiveness of air jets depends upon variation in flash, which is in turn a function of the casting machine, die condition and of die and metal temperatures. Operators can observe flash and remove it. A robot, however, lacks vision. Consequently dies must be of a higher quality for automatic operation, and continuously maintained in top-class condition.

There are, however, many advantages to this application of the robot. First, savings are high because in most cases two operators are eliminated. Second, material handling and storage space are reduced. Third, production yields are higher than manual operations which involve immediate trimming following casting because of the consistency of robot operation. This consistency produces more uniform castings, less rejects, and, with good dies, fewer trimming press problems caused by casting shrinkage and warpage.

5: Unloading a die casting machine with die care performed by robot

The robot's versatility and repeatability enables it to blow off flash and apply lubricant more effectively than the manual operator or any other automatic unloading technique. This is particularly useful in aluminum die casting. Proper die care accounts for an appreciable percentage of the total casting cycle time and therefore has a great influence on the production rate. Also, the higher temperature alloys require more meticulous die care. These factors coupled with short runs make it difficult and sometimes impractical to employ automatic die care methods which require set-up time of themselves. The ease with which the robot can be taught permits the proper air cleaning and lubrication patterns for a particular die to be achieved readily.

In this application a typical sequence is as follows: the robot's function begins with it in a 'ready' position, waiting for the die casting machine to open. If the machine does not open, the robot does not move but an alarm is sounded as part of the program. When the machine opens, the robot enters the die and signals the machine to eject the casting. It then grips the casting and places it on a chute which carries it to an inspection table. As a safety precaution the part must touch an electrical limit switch during its passage down the chute before the robot releases its grip on it. After removal of the casting the robot activates water sprays used to cool the casting die.

In its next operation the robot grasps a spray gun, enters the die, activates the gun and lubricates all die surfaces. It sprays in a precise, programmed pattern reaching all areas of the die, withdraws and signals the machine controls to close for the next cycle. The big advantage of the robot is that it lubricates a die more consistently than does a man. The lubrication sequence is essential for die life and quality castings. The robot performs this sequence correctly every time. At the end of the cycle the robot puts down the gun, returns to the start position and waits for the machine to open again.

Since some dies do not require care after every shot, a robot in this application can be commanded to perform the die care portion of its program only after so many shots have been completed. This intermittent operation is accomplished by setting a selector switch and its frequency can be altered in the same manner during any run. Figure 10.8 depicts a Unimate robot shown carrying out die care.

6: Load insert into die casting machine and unload casting

Inserts of materials that are different from the casting alloy can be joined to die castings during the casting operation. For example a steel shaft required for strength and hardness can be inserted in a die cast gear wheel. Similarly, when bearing requirements are too severe for the casting alloy, a sleeve of bearing material may require to be inserted.

Figure 10.8 *Robot engaged in die care procedures*

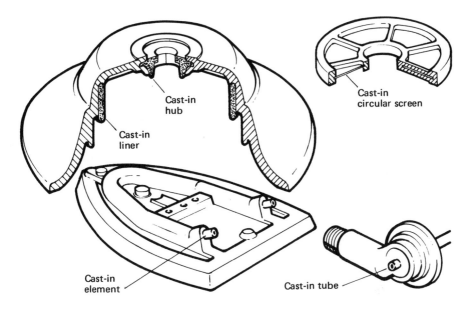

Cast-in hub

Cast-in circular screen

Cast-in liner

Cast-in element

Cast-in tube

Figure 10.9 *Die casting capability extended by cast-in inserts*

This can often be incorporated during the casting process. Magnetic materials as well as non-conductive materials such as ceramic, plastic or glass can also be inserted. Typical inserts are shown in Figure 10.9.

The robot can be programmed to place the insert in the correct position inside the dies, usually on some form of attachment or in a hole provided to retain the insert while the casting process is completed. After positioning the insert, the robot withdraws and actuates an interlock device which commands the platens to close. After the shot, and allowing for an appropriate cooling period, the platens open and the robot continues with a typical sequence such as are described in applications 1 through 5. In one such application, a 2000B Unimate robot is being used to service a large die casting machine producing electric motor end housings. In this job the robot replaces 1½ workers per shift, and it removes the hazard of working between the dies of the die casting machine and trim press. Productivity has been increased by over 10%.

The motor housings are produced in a two-cavity mold which contains two inserts that serve as shaft bearing mountings. The robot picks up two inserts from gravity magazines, swings to the die casting machine, unloads the previous shot by grasping the biscuit, and then loads the inserts while holding the shot. Then the casting is placed in the quench tank and a previously quenched part is transferred to the trim press for trim and disposal. Press open time is about 8 seconds to unload the shot and load the two inserts. The production rate is about 160 shots per hour.

The Unimate in this case uses a special double hand. The hand includes the biscuit gripper and two expanding mandrel devices for holding the inserts. The hand swivels 180 degrees between unloading the shot and loading the inserts. Operation of the insert holders can be programmed independently of the biscuit gripper. A variety of insert mandrels can be provided to accommodate different inserts, while their construction, of non-magnetic stainless steel will allow magnetic retention of the inserts within the die. Figure 10.10 illustrates both the plant layout and a specialized hand design for using robots to position inserts into the die.

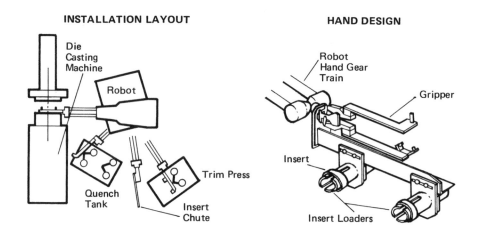

Figure 10.10 *Equipment layout and hand design for insert positioning*

More recently some very advanced insert developments have been successfully accomplished. Notable among these is tooling to position cylinder liners in a modern V-6 automobile engine block. This requires the robot to lift a heavy piece of tooling into the dies, whereupon actuators within the tooling push the liners into position. The robot then withdraws and the cycle proceeds, during which time the cooling is replenished with liners. In view of the large size of the casting and the lengthy positioning process which would normally have to be carried out by a manual operator under very adverse conditions, the new application is increasing production rates even at its early stage of introduction.

Further considerations for robot die casting

While the above section describes a number of ways in which robots can be employed in die casting, the finer points in any installation will best be determined by the die caster, usually after some experience with different methods of approach. There are, however, several common denominators involved in achieving optimum performance, whatever the mode used. Perhaps the best way to consider these is to pose certain questions and suggest that the die caster ask them of himself. For example:

1 What casting times are involved? These times will exert an influence on the amount of work that can be done by a robot during the die-closed time while still meeting the required production rates.

2 Is quenching necessary? If so, how should it be done? In some instances the casting must be fully immersed. In others a curtain or spray quench is adequate. Whatever is decided upon, the equipment should be arranged so that the required robot motions are minimized in the interest of short time cycles.

3 Is die care to be performed by the robot or by some other means? Among other things, this decision will influence the type of robot grippers or fingers which must be used.

4 Is there a tendency for the casting to fall off the ejector pins? The robot has more than adequate power to remove the casting from die. Of greater concern is the possibility of the casting falling off the pins. Falling can usually be avoided by slight alterations to the die.

5 How is the casting to be gripped? The conical shape of the zinc die casting sprues is ideal for firm gripping. The biscuit on aluminum castings is sometimes difficult to hold, and gripping a portion of a runner may be better. Interferences due to multi-parts dies and core pullers may dictate the finger shape and how the casting is gripped. Protection of exposed and finished surfaces must be considered.

6 Is separating or trimming involved? Efficient equipment arrangement is vital if it is. Casting distortion and warpage may require beefed-up runners. Precise part orientation required in the trim die may dictate the way in which the part is grasped.

7 How is successful casting removal to be detected? In some instances sensing any one portion of the casting is adequate for die protection. In others, many areas must be sensed. When unit dies are used it is desirable to be able to blank some sensors when a cavity portion is unused.

8 What kind of sensors should be used? Limit switches are simple but not entirely reliable and do not always lend themselves to sensing many areas at once. However they are inexpensive.

Infrared, pyrometric and photoelectric devices offer a high degree of discrimination and can sense either large or small areas. Clusters yielding one go or no-go output are readily available, and they usually permit greater flexibility in robot programming compared with the use of mechanically-operated switches. Their initial cost is considerably greater than limit switches but this is more than offset by their life and reliability.

9 What kind of interlocks should be considered? There is perhaps no other single requirement which is more critical to a successful installation. Yet the interlocks are likely to be grossly inadequate and unreliable. The human operator can and does overcome these deficiencies, but without sight or reasoning power the robot is not so forgiving. The robot manufacturer has devoted considerable attention to the interface requirements so that efficient, reliable operation can be achieved easily. It remains for the die caster to take advantage of this facility. Sometimes rudimentary interlocking controls are the cause of the problem, but the other extreme of an excessively complex system can be just as troublesome. The starting point in appraising the interface and machinery control requirement is to ensure that there are no provisions which exist only to overcome deficiencies that can be corrected. Every element in a system has its own reliability factor which lowers overall reliability. The fewer the elements, the higher the net reliability.

Once it is determined that the controls of the die casting machine (and any other machinery) are as efficient and reliable as possible it must be decided what information should be exchanged between robot and machines. As an integral part of the program, the robot can be taught to start the cycle of the die casting machine or trimming press at the appropriate time and to wait until certain external conditions are satisfied before proceedings with its own cycle. Typically the robot is ready to remove the casting as soon as it is told that platens are fully open, cores retracted, ejector pins advanced and so on. It removes the casting and presents it to the sensors which, together with a robot signal, will determine a 'go' condition whereupon the next die casting machine cycle is started and the robot continues its own cycle. Further refinements such as synchronizing ejector pin action with gripping of the part by the robot, activating die care cycles, inspecting parts and automatically disposing of rejects and the like, all lie within the inherent capabilities of the robot.

10 What safety provisions are needed? Two considerations are involved in applying safety techniques. First and foremost is protection of life and limb. The second is protection of costly equipment.

In the first instance, many standard techniques used throughout industry can be applied. However, an additional factor is introduced. In installations where the full sweep of the robot's arm is not used, and safety fences are placed inside its reach in the interest of conserving floor space, a combination of mechanical and electrical safety limits, internal or external to the robot can easily be arranged. Protection of equipment is usually implemented by appropriate interlocks such as those already discussed. However it may be desirable to limit certain motions of the robot as an added safety precaution.

Chapter 11

Spot welding applications

Welding is the process of joining metals by fusing them together. This is unlike soldering or brazing where joints are made by adhesion between two surfaces alloyed with a metal of a lower melting point, such as lead, tin or silver.

For hundreds of years blacksmiths have welded wrought iron by heating the pieces to be joined almost to melting point and then hammering them together on an anvil until they become virtually one piece. In modern welding the heat required to cause the metal to fuse is provided by gas torches, electric arcs (see Chapter 12) or by the passage of an electric current through the metals at the point where they are to be joined.

Outline of spot welding operation

In spot welding, as its name implies, metal pieces are joined at a number of small localized areas or spots. This is accomplished by passing a large electric current at low voltage through the metal at each point to be welded. Sheets, rather than large metal structures, are suitable for spot welding since more massive pieces will conduct the heat away much too rapidly, making it impossible to raise the temperature of the part sufficiently for fusion to occur.

The heat in spot welding arises from work done by the electric current in overcoming the small, but finite, resistance offered by the metals to be welded. In fact, the process is sometimes referred to as resistance welding. For a given level of current, the higher the resistance encountered, the greater the heat generated. In practice, the materials being joined are clamped between copper or copper-alloy electrodes which conduct the welding current to the site of the weld. As current flows through the workpieces from one electrode to the other, localized heat is produced in the column of material directly underneath the contacting area of the electrodes. If this heat is sufficient to fuse the material, then welding takes place at this 'spot'.

The pressure exerted by the electrodes on the two surfaces to be joined is important since it to some extent controls the resistance of the conducting path. As pressure is increased, the resistance across the junction will diminish. This means that a higher current is needed to ensure sufficient heat generation for fusion to occur. On the other hand, too little pressure results in an unduly high resistance path so that the area under the electrodes may burn away due to excessive heating. It follows that there is a critical relationship between the current through the materials and the time during which this current flows; a high current flowing for a short period will generate as much heat as a smaller current flowing for an appropriately longer time. From this it will be seen that there is a need to exercise a rigid control over the welding cycle, and specifically to ensure that the level of current, the pressure between electrodes and the time during which current flows are all correctly selected to suit the particular conditions imposed by the work in hand.

Welding sequence

A typical spot welding sequence comprises the following operations:

1: Squeeze .... The two surfaces are held together in contact with the electrodes which exert a force of 800 to 1000 lbs per square inch.

2: Weld ...... Current is turned on and flows through the material. Heat is generated in the vicinity of the electrodes.

3: Hold ...... The tips of the gun are kept closed long enough for the weld to cool, usually assisted by water circulation through the electrodes.

4: Off .... The machine is rested until the next operation. In order to avoid overheating of the welding equipment, the 'off' cycle is governed by the welding control unit which programs an interval long enough for cooling to occur.

Modern spot welding machines are equipped with automatic controls which can be set up to optimize this sequence of events for the particular work to be done. For the production of high-quality welds the machine must be capable of accurate and repeated production of an electric current impulse of a predetermined shape and duration, and must be able to maintain the same rate of heat absorption (and subsequent dissipation) by the electrodes at each successive weld.

Since the bulk of spot welding machines used in industry are powered by alternating current supplied from a power line, the frequency of this supply (typically 60 Hertz per second) is utilized as a timing source for the control system.

To give an example, the machine might be programmed to squeeze for 8 complete cycles, to weld during a 9-cycle period and to hold for a single cycle, making the entire sequence last for 18 complete cycles or

18/60 second. Considerable variations in sequence times are available with modern machines to provide maximum flexibility of operation.

Suitable materials

When the material being joined is thick, or when certain metals are used which do not conduct adequately, a sequence step known as double heat can be programmed. This is accomplished by the controller increasing the weld step in the program to the appropriate time required.

Spot welding is applicable only to electrical conductors — metals for all practical purposes. Some metals such as aluminum and copper are rarely suitable since their electrical resistance is so low that they would require excessively large currents to raise their temperature sufficiently for fusion to occur. In general, ferrous metals are those most suited to spot welding. The most obvious applications are in the manufacture of domestic appliances, particularly appliance cases, automobile bodies and sheet metal fabrications of all types.

Design of welding guns

The tong-shaped assembly of electrodes used in spot welding is known as the gun. In one type, the C-gun, one electrode is fixed while the other is free to move. The fixed electrode is brought up to make contact with one side of the joint to be welded and the remaining electrode is then moved to sandwich the work between the two conductors. In another type, the pinch gun, the electrodes are joined by a hinge so that both are free to move. This gun is provided with an equalizing mechanism so that when one electrode touches the work-piece, it comes to rest, whereupon the other electrode moves rapidly to close on the work. The tong-like design may need to be considerably elongated in guns which have to make welds across a wide section (such as the central area of a car-body floor). Such guns are said to have a large throat clearance; up to 30 inches may be necessary for certain applications.

In some cases a so-called back-up or retract gun is used. This type of gun has an additional stroke which opens the tips more widely to permit it to pass over flanges or other thicker metal sections. The choice of gun is very important. If at all possible the need to change gun types on a welding operation should be avoided in view of the time required. This will be discussed later in connection with an actual application. Figure 11.1 shows a typical gun used in automotive body fabrication.

The electrodes are subjected to intensive wear and tear since they operate continuously through high temperature cycles which causes them to burn away. To extend the life of the guns, replaceable tips are

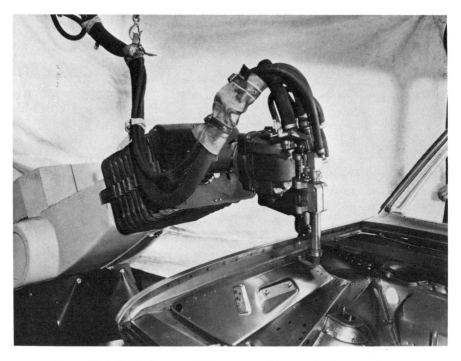

Figure 11.1 *Typical spot welding gun used in auto body manufacture*

fitted, and it is common practice to dress the tips manually using a file from time to time to improve their shape. Unless this is done their contact resistance will alter and the quality of the weld will suffer.

Since the currents used in spot welding can be as much as 1500 amperes heavy cables are required to connect the electrodes to the supply. These must be terminated in such a way that movement of the gun is not unduly impeded, especially when it has to be moved around into difficult corners or inaccessible places. To help bear the weight of these cables the gun assembly is incorporated in a welding head, a framework from which the bulk of the weight is taken by an overhead suspension system but which can be moved around by the operator. These heads can weigh as much as 200 lbs unsuspended.

To complicate the welding head further, it is almost always essential to provide water cooling of the electrodes for production welding systems. Without this serious overheating problems would occur which would significantly reduce the life of the electrodes.

The welding sequence is so short that the bulk of the time taken to complete a job involving several welds is in the movement of the gun between spots. Because of the difficulty of handling the unwieldy gun for long periods, and the fact that the operator needs to use judgement in deciding where to position a weld, spot welding is not a very precise operation. This leads to the question whether the process can be automated to remove some of the drudgery from the job and to improve accuracy by removing the human element. When the product line is unlikely to change significantly over a long period, then the use of a special purpose automation system would be justified. This might be the case, for example, in a box for a refrigerator which might change very little over a period of years. However, when the line is subject to frequent change, this solution is not economical. This situation prevails in the automotive industry where models traditionally change every year and where typically three or four different body styles may be under construction simultaneously on the same line. In these situations the robot welder comes very much into its own.

Robots in spot welding

Over the past decade the main application of industrial robots throughout the world has been in the field of automotive spot welding. It is by no means a simple process; on the contrary it is complex. Yet in many ways the characteristics of a robot are ideally suited to this application. The entry of the robot into this domain came about as the result of painstaking development work on the part of robot manufacturers and cooperation from the car makers — the most profit conscious industry of all! It happened in the following way. About 15 years ago in the car industry, body parts, some of them large sub-assemblies, were held in clamping jigs and tacked together by manual operators using multiple welding guns. The unit would then be finally spot welded by operators who would position the welds using their own judgement. In 1966 the first steps were taken to use a robot to guide the welding tongs, linking its control unit with external signals which commanded the robot when to squeeze, weld, hold, etc. By 1969 General Motors had installed a line of 26 Unimates for spot-welding car bodies, and this was followed in 1970 by Daimler Benz in Europe who adopted a Unimate for a body-side spot-welding operation.

From those beginnings, the use of robots in this application has mushroomed, and so important is this process in the industry that it has been necessary for the car manufacturers to consider the capabilities and limitations of the robot in designing the production line itself. Step by step the introduction of robots into the car-body spot weld process has continued until more than 1200 such machines are now in use.

To take its place on an automotive production line the robot must be able to remember several different body styles. This is a difficult task, since cars are designed from a stylistic viewpoint, with the result that the geometry often gives little help to the robot in establishing reference points such as it might encounter in more functional shapes like appliance housings. Having absorbed the necessary information in its memory system, the robot must, on command, initiate the appropriate program to suit the type of body to be built and then carry out a series of complicated manipulations using a heavy spot welding gun. It must perform these tasks with speed, accuracy and reliability in an industry producing up to 80 car bodies per hour on the line.

This is a job for a high technology robot such as those available from ASEA, Cincinnati, Kuka, Comau, Kawasaki and Unimation.

All of these models are to be found in this application today. They often work a three-shift program, although the third shift is usually a shortened one during which maintenance can be carried out on the line as a whole, not just on the robots. The repeatability and positional accuracy which the robots can achieve provides a much more consistent performance than is obtained with human operators. In fact body strength can be obtained by specifying fewer welds (in the correct locations) than would be the case with a human operator who could not be relied upon to produce the best pattern of welds.

Much of the work currently done by robots is in the area of respotting. This is a technique where an initially tacked-together body is finally welded by a group of robots on the production line. The group will be under the control of a single supervising computer which will signal the arrival on-line of a particular body style and cause the robots to switch to the appropriate program in their memory systems. Various interlocks and sensing devices can easily be incorporated to ensure that the robots know what they are expected to make, when to start, stop, etc. Robots thrive on the sort of commands which are inherent in spot welding cycles.

All major car manufacturers are now actively using robots on their production lines. In fact two manufacturers, Volkswagen and Renault, have developed and built their own robots with the specific objective of doing spot welding and little else.

In the following text a brief description of some of the factors which must be considered in specifying a welding installation will serve to define the process.

Planning a robot spot welding line

In planning to use a robot for spot welding with a portable gun, a method of estimating the time required to perform a specific group of welds is needed to determine the number of robots and individual work stations which will be needed.

The initial determination to be made is the number of different guns required for the whole task. Assuming that it is preferable not to have more than one gun for each robot, because of the lost working time involved in changing the guns, the initial sorting out that will be part of the weld study will be to group together all those welds which can be done by the same gun.

Some discretion will be necessary here to avoid the choice of one single large clumsy double-acting gun for all welds and thus suffer the penalty of low maneuverability and speed.

When this preliminary selection has been made, and an initial attempt made to assign specific welds to individual robots, then it is useful to have some standard method of estimating the total cycle time. To accomplish this, application engineers have established certain definitive movements or operations which will be required in spot welding applications. These can be timed and used as data to calculate, to a reasonable degree of accuracy, the time for the whole weld cycle. Such generalized motions are:

1 Moving from an at rest position to the proximity of the first weld. (IN)
2 After positioning at the work area, adjusting the gun attitude to be properly normal to the metal (or flange). (MINOR ZONE CHANGE)
3 Performing a group of sequential welds in which the distance between welds is small and no significant change in the attitude of the gun is required. (ZONE GROUP WELDS)
4 After performing a group of evenly spaced welds, shifting or skipping along the flange to a second group of evenly spaced welds. (MINOR ZONE CHANGE)
5 After performing the first group of welds, the flange changes its plane of orientation due to the necessarily convoluted shape of the metal. (MINOR ZONE CHANGE)
6 Withdrawing the gun from the flange, removing it completely from the work in order to move to an entirely different area, and orienting the gun around the new flange. (MAJOR ZONE CHANGE)
7 After all welds are completed, withdrawing the weld gun to a clear position so that the metal can be indexed or transferred from the work station. (OUT)

To these times must be added the time taken to accomplish each individual weld. The robot will be standing still while the gun tips are closed on the metal for a period equal to squeeze + weld + hold, all of which are accurately known since they are controlled by the welding controller and timed using the 60 Hz supply as previously described.

Figures 11.2 and 11.3 illustrate some minor zone changes, while Figure 11.4 shows a profile of an automobile car body section requiring a major zone change.

The net result of the planning operation will be a robot installation of a type shown in Figure 11.5 where 13 robots are utilized in 7 separate

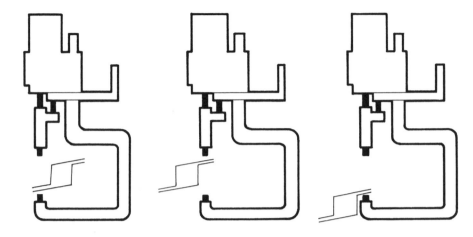

Figure 11.2 *Spot weld ~g: simple minor zone change* (only two motions: → and ↑)

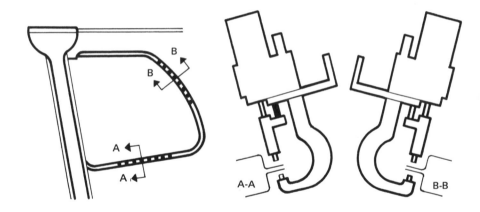

Figure 11.3 *Spot welding: complex minor zone change with rotation of gun and change in angle*

welding stations along the production line.

The best way to show the capabilities of robots in a spot welding application is perhaps to describe an actual installation. The Volvo plant illustrates the present state of the art in this regard.

The Volvo spot welding installation

In Gothenburg, Sweden, Volvo is using the largest and most sophisticated automatic welding line in the world. A line which previously required a workforce of 67 has been replaced, at a cost of $6 million, by a robotized line serviced by only a handful of key staff. The line is

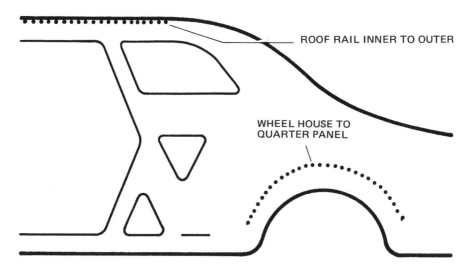

ROOF RAIL INNER TO OUTER

WHEEL HOUSE TO
QUARTER PANEL

Figure 11.4 *Spot welding: major complex zone change on a typical
auto-body weld job*

Note: The wheel house welds are underneath. The gun must be dropped down and
withdrawn; then lifted and turned around, then dropped down and engaged with
the roof rail. The involvement in disengagement with the wheel house makes this a
COMPLEX zone change.

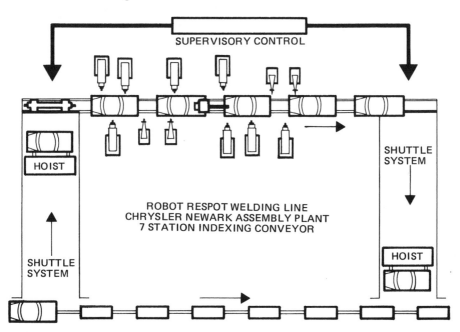

SUPERVISORY CONTROL

SHUTTLE
SYSTEM

HOIST

ROBOT RESPOT WELDING LINE
CHRYSLER NEWARK ASSEMBLY PLANT
7 STATION INDEXING CONVEYOR

SHUTTLE
SYSTEM

HOIST

Figure 11.5 *Typical robot grouping on spot welding line*

scheduled to produce 50 car bodies an hour in a mix of 2-door and 4-door models, and the aim is to improve the working environment and produce a more consistent product. The installation was designed by an Italian company which already had designed and installed robot welders in the Fiat works in Italy, Fiat being Europe's biggest user of robots among the car manufacturing fraternity.

This new line introduces several novel features and is the most advanced yet to be built. It consists of ten work stations, connected by a conveyor system. The car bodies in various states of completion are at all times mounted on pallets. The line is on two levels, the upper one being for welding and the lower for returning the pallets to the starting point. An interesting mechanical feature is that the welding line is indexed whereas the return line moves constantly at about 50 feet per minute. This allows pallets to congregate around the starting point, making them available as required. However the subject of most relevance to this application note is the role of the robots. This is best described by outlining what occurs at each of the ten stations along the line.

Station 1. This is the loading station at which the body sections are loaded on to a pallet to be conveyed along the line. No robots are used.

Station 2. At this station the body parts are held in position by swinging 'gates' while they are tacked together by six robots operating horizontally (Unimate 4000B) and one operating vertically (Unimate 2100). A total of 128 spot welds are carried out at this station.

Station 3. This is the first of two respotting stations at which the tacked body is finally welded. Here, three robots perform 82 spot welds, two operating horizontally (Unimate 4000B) and one vertically (Unimate 2100C).

Station 4. At this, the second of the respotting stations, 138 spot welds are carried out by five robots. (Four are horizontal Unimate 4000B models and the other is a vertical Unimate 2100C).

Station 5. This station illustrates the way in which conventional automation and robots complement each other on this line. The controller here selects a roof from two available types, normal and hatch. It places the roof on the car body shell with sufficient precision to permit the robot installed here (Unimate 2100C) to fix the roof using 14 spot welds.

Station 6. This station is reserved for final roof fixing plus a general check of the body at this stage. A robot finalizes the roof position by spot welding in the drip channel, while a manual operator loads tie-plates into an automatic transfer machine which then places them into the correct position for them to be welded into the body. Six robots perform this operation by making 176 welds. Four are Unimate 4000B and two are Unimate 2000C).

Station 7. No robots are used at this station. Instead, welding guns guided by templates which match the roof configuration produce overlapping spot welds to make a seam. There are four of these guns which each make 60 welds while moving at about 7 feet per minute.

Station 8. This is the point where the final respot is carried out to complete the welding operation on the body. Five robots are employed here (four Unimate 4000B and one Unimate 2100C) and a total of 156 welds are made.

Station 9. This station is purely for back-up and checking. If any robot trouble develops, manual labor can be provided at this point to complete the body.

Station 10. At this position the body is unloaded and the pallet returns by the lower path to the starting point.

The above is typical of the use of robots in spot welding, yet the application is relatively simple. Only two body styles are involved so that the robots being employed are by no means operating at the limit of their memory capabilities.

Co-ordination of spot welding robots

When a large group of robots is assembled together on a line, their control can be centralized. Furthermore, if the plant using them already uses advanced computer techniques in its production scheduling, the same computer can be used to provide command signals and to call up appropriate programs to cater for varying body styles along the line. The computer monitors and controls the whole system, while simultaneously recording production statistics and checking the welds for quality.

The use of swinging gates to introduce the body sections is becoming widespread in the industry. What this particular installation does is to do away with the usual multi-spot welding jigs which have always been used at the tacking stage. The overlapping welds made by contour-guided non-robotized guns is also novel, indicating that robots, conventional automation and even human operators can work in complete harmony if a systems approach is adopted from the outset!

In fact the combined onset of a number of robots which occurs when they are signalled that a body has come on to station is very reminiscent of the action taken by a group of manual workers assigned to a similar task in pre-robot days, each robot concentrating on its own particular job and ignoring what is going on alongside.

While robots have demonstrated their capabilities on a line such as is described above, they can clearly do other work in support of the operation where this can be seen to be economically justified. Although not yet perfected, it is likely that electrode tip dressing will eventually be handled by robots during any downtime or waiting time which they have at their disposal.

The spot welding of automobile bodies by robots is a difficult enough task, requiring six-axis articulation to accomplish. On an indexed production line, the parts being welded are at rest during the actual welding operation. Generally speaking, however, automotive manufacturers have long worked with continuously moving lines, and this poses special problems for the robot which must now be programmed to move during the welding sequences in order to accommodate the steady movement down the line of the body under construction. This is something which a human operator can accomplish almost intuitively, but with a robot the extra memory and control requirements are considerable if this feature is to be added.

Chapter 12

Arc welding applications

The spot welding process described in Chapter 11 is not suited to all welds, especially long-path joints needing a gas-tight seal between the two surfaces being joined. However an alternative electric welding technique exists for such situations in the form of electric arc welding. In this process the heat required to fuse the metal surfaces together is derived from an electric arc. This is no more than a sustained 'spark' or electrical discharge between two terminals, which in this case are the work and a metal welding electrode.

When the arc is struck, the temperature in the vicinity rises rapidly to as much as 6500 degrees Fahrenheit. At such high temperatures, a small pool of molten metal forms in the work, and the end of the electrode also melts to contribute additional metal to the pool. Obviously the electrode material must be electrically conducting and compatible with the metal forming the part being welded. A typical electrode is in the form of a metal wire, continuously fed at the correct rate to replace electrode material consumed by the welding process.

Arc welding process

Originally electric arc welding was performed using a carbon rod electrode which sustained the discharge but supplied no material to the weld. Indeed it was usually of importance that the amount of carbon particles finding their way into the weld be minimized since their presence could impair the quality of the join. By playing the arc back and forth across the area to be welded, the workpiece was maintained in a molten condition, and if additional metal was needed it was supplied by means of a separate filler rod.

The carbon arc method has now been largely abandoned in favor of more sophisticated procedures. One of these is to use a metal rod coated with a flux material as the electrode. The welder inserts the rod into a holder, and the heat from the arc melts the flux which flows freely over the area of the weld, preventing oxidation. When the rod is used up, another is inserted. The method is simple and suited to portable welding units which can be taken to the site of the weld.

A more advanced technique is the so-called Heliarc welding process in which a single electrode such as tungsten is again used solely to provide and sustain the arc, as in the carbon electrode method. Since tungsten has a high melting point, it does not melt and therefore provides no material to the weld. A filler rod must again be utilized to provide any 'make-up' material required.

What distinguishes the process from the carbon arc method is the fact that during the weld, the area surrounding the arc is flooded with an inert gas such as helium (hence the name Heliarc,although argon is also used), and this gas continuously protects the weld from the atmosphere to inhibit oxidation. The method is useful in welding metal such as aluminum and its alloys, copper, magnesium and stainless steel, and is known as the Tungsten Inert Gas process (TIG).

In factory welding installations used on production-line fabrication the most commonly used method is currently the Metal Inert Gas technique (MIG), also known as Gas Metal Arc Welding (GMAW). It uses the shielding effect of inert gases found in TIG, combined with a continuously fed metallic electrode, usually in the form of a wire coiled on a drum. The electrode material is selected for the particular weld since it provides filler metal.

Modern electric arc welding equipment comprises a gas line (flexible tubing) through which the electrode wire is threaded. The line and the wire terminate in a welding gun, the wire protruding to form a convenient tip with which to trace out the line of the weld. Control equipment propels the wire through the gas tube and governs its rate of travel. The voltage applied between the work and the electrode is monitored and used to sense the weld conditions so that control signals can be generated to make the operation as automatic as possible. Figure 12.1 illustrates a typical electric welding gun.

Whereas alternating current is almost always used in spot welding, the electric arc method operates from direct current, usually in the range 100 to 200 amperes at 10 to 30 volts. Conventionally the positive side of the supply is connected to the electrode, and the work goes to the negative terminal.

When a manual operator uses arc-welding equipment he will clip the work to one side of the supply, the other side being already connected through the control system of the machine to the electrode. The electrode is then simply a wire, and with power on, the operator touches this wire on to the work at the point where welding is to commence. This virtually short-circuits the supply, and generates a good deal of local sparking and heat generation.

By withdrawing the electrode an appropriate distance from the work, the operator is able to initiate and maintain an arc discharge between the electrode and the work. Human judgement is required throughout the weld, to position the electrode at an optimum distance from the work, and to move it along the line of the weld at an even rate to

Figure 12.1 *Typical robot arc welder*

produce good molten metal flow between the surfaces and a proper
build-up of metal without burning away the workpiece by generating
too much local heat.

Sometimes the electrode will stick to the work, and the effect is to
cause the voltage between the electrode and the work to fall to zero.
The control elements within the power unit are designed to protect the
supply from damage caused by excessive current drain. Meanwhile the
operator must free the electrode and re-position the tip to get the arc
going again. It will be evident that the whole procedure is subject to
human variations and error so that no two welds, even those performed
by the same operator, can ever be totally alike.

Furthermore, the electric arc process creates an unpleasant working
environment. Considerable ultra-violet radiation is present in the
intense glare from the arc itself, requiring the operator to wear almost
opaque goggles or a protective mask to filter out these dangerous
radiations and prevent them from damaging the eyes. Sparks tend to
fly around the shop, and smoke is generated, so it is little surprise that
it is difficult to find welders who are prepared to work a full shift under
such conditions and who are at the same time capable of consistent,
repeatable results. Once more, enter the robot to take over, unaffected

by the environmental problems and virtually incapable of any deviation from a well-defined task that it has once been taught.

Robots in arc welding

Good electric arc welding requires close control of the welding gun along the weld path, both in position and speed. An experienced welder selects the best angle of attack, the right dwell-time and the correct supply voltage to produce good fusion of the surfaces, and the result will be seen in the uniformity of the finished work and the absence of unnecessary metal build-up or blow-holes, however complex the job. Inevitably a certain element of 'green thumb' operation defying precise description, must characterize the work of really top-class welders. Lacking this feel for the job, it at first sight seems an impossible task for robots, but in fact it is a classic opportunity for them to show what they can accomplish.

Industrial robots were conceived as replacements for human operators, and their justification has always included that should they fail on the job, a human operator can always step in to maintain production while the robot is repaired.

Assuming that a job is being done efficiently by a human operator, then when a robot takes over it should ape the operator's movements as closely as possible. This is perhaps especially true in the case of arc welding. In an ideal situation, a robot would be taught an arc welding sequence by having an expert welder literally take it by the hand and lead it through a weld cycle on an actual part. In the grasp of the robot would be a conventional welding gun, and as the operator guided it along the contours of the weld, all the positional information would be recorded in the program electronically.

Once taught, the robot should be capable of repeating the operation for as long as required, achieving a greater consistency of weld than is possible with a human operator. There is more to it than that, of course. But simply stated, the job of the robot welder is to place the welding gun in the right place at the right time, throughout whatever path the weld requires. This implies the need for both positional as well as speed control. The speed of the gun will seldom be constant over the entire path, but instead will be subject to variations as the gun follows the contours of the weld, some of which may involve making tight turns to be made or welding to take place in inaccessible places.

Most arc welding jobs require a robot with but five articulations because the weld gun is a symmetrical tool.

An attractive feature of robot welding lies in the fact that it represents true human-replacement, and as such handles perfectly standard arc-welding equipment to do the job and does not demand much in the way of specialized equipment. All that is needed is a fitment to mount the welding gun, attached to its long gas lines and wire feed mechanisms, on to the robot's hand. All other equipment associated with the arc welder will then operate in normal fashion.

The robot's job is mainly to guide the gun around the programmed path and to signal when it is on station and ready to proceed. The welding unit controller does the rest.

Figure 12.2 illustrates a typical weld sequence in which command signals are provided jointly by the robot and the welding equipment control circuitry. The sequence is:

1 Inert gas is turned on to flood the work area (Pre-flow period)
2 Wire feed commences, power supply comes on (Weld period)
3 Wire feed ceases, surplus wire burns off until gap between electrode and work is no longer small enough to sustain the arc (Burn-back period)
4 Power goes off while gas remains on (Post flow period) to inhibit oxidation by hot surfaces
5 Gas goes off — weld sequence completed.

The robot, on a signal that the part is ready, moves to the first step in its program, usually the point where welding is to commence. The robot then initiates a signal to the welding unit controller and step 1 (above), the gas flow, commences.

Step 2 is initiated by the welding controller, and the same signal commands the robot to start welding as the arc strikes. At the

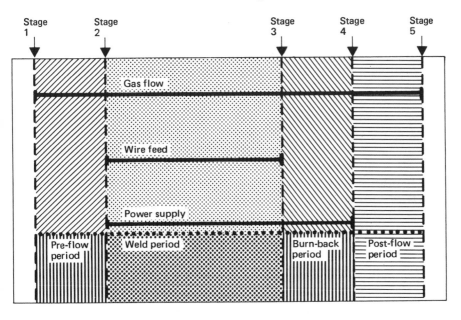

Figure 12.2 *Electric arc welding: typical weld sequence*

completion of the weld, the robot signals the welding controller to initiate steps 3 through 5. The robot then waits for a signal telling it that the next part is ready, and it moves to its first program position and the whole sequence is repeated.

It might be argued that none of the above requires a robot, since a less sophisticated machine could be constructed to do the job by the use of cams, limit switches etc. This is indeed true, and such welding machines have not only been built — they are in use. They are, however, limited to runs where the product line is not subject to change. If changes are made to the parts to be welded, extensive modifications must be made to the machine which are both costly and time-consuming. With a robot, the insertion of a new program, preferably recorded by an expert welder, will equip the willing worker immediately for its new job.

Programming the robot

It was mentioned earlier that the true role of arc welding is to make long joins either for increased strength or for gas or liquid-tight seals. It follows that the path followed by the robot in making such a weld will be long and narrow, and due to the complexity of the part, it may also be strangely contoured. The best type of robot to follow such a path would appear to be some form of continuous path machine. However, to control the moving path of a robot still presents technical problems. If it is required to move from one fixed point to another, it is relatively easy to arrange this, and the robot will take the best route for itself, usually anything but a straight line. To dictate the precise path followed between the two points is very much more difficult. Nevertheless this is exactly what must be done if a robot welder is to be able to follow any welding contour accurately enough to emulate or even improve on the job done by a human operator.

One way out of the predicament which is frequently used is to divide the path into a large number of small, equal steps. This can be done on the actual part to be welded by positioning around the line of the weld an adhesive strip (see Figure 12.3) on which are printed small increments which are numbered for easy reference. The operator can now concentrate on each small step separately, and optimize the path of the welding gun over that step, meanwhile recording the information in the memory of the robot. This assumes that the robot will follow a

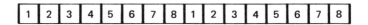

Figure 12.3 *Section of tape used to break down a welding contour into small equal steps*

predictable path, preferably a straight line, between the two points, but we have already commented that this is no simple matter to arrange. However by using an interpolation technique a fair approximation to a straight line can be achieved — certainly close enough to meet the specified tolerances of a typical weld. The advent of cheaper computing through the use of microprocessors can be expected to have a great impact on path control on robots in the future. However a program recorded in the way described above is quite satisfactory and has certain inherent advantages. It can be played back over and over again and individual steps in it re-programmed to refine the path until it meets with the approval of the operator. Then all that needs to be done is to decide on the speed of the weld and to set up the robot to provide this as part of the program. The machine is then equipped to go into production.

Assume for a moment that it is possible for a welder to take hold of a robot arm fitted with a welding gun and to carry out an actual welding sequence, all aspects of which in both position and speed can be recorded into the robot memory and played back in real time. Then many trial runs may be required before an acceptable program emerges. This can be expensive in parts, whereas breaking down the program into separate steps is more time-consuming but more likely to produce a good program first time out. Both approaches are used effectively, but the point-to-point method is probably superior on complex jobs.

Choice of robots for arc welding

It has been stressed that despite the apparent simplicity involved in moving a welding gun around a given path, arc welding, even by robots, is a difficult task and a challenge to the robot designer.

The use of a large machine like the Unimate for arc welding may be justified from the point of view of its spatial command, but its ability to handle heavy loads is generally wasted in this application since welding guns typically do not weigh very much unless they are of the heavy water-cooled type used on massive work. This raises the question whether a hydraulically operated machine is best, or whether an electrically driven one might not do better.

In many applications there is sound economic justification to opt for electrical drives. Slow speed and low pay-load make electric drives less costly to operate than hydraulic drives. Hydraulics excels for high speed and heavy loads.

Real time programming has the advantage of using an operator in the welding cycle but if a mistake is made, the program must be started over from the beginning. Also, carrying the robot arm around while programming can hinder the operator who is used only to carrying a torch.

Point to point programming with computer interpolation allows for easy program touch-up until an optimum is achieved. Both methods have their devotees.

Out there in the shops are machines made by Unimation, Kawasaki, ASEA, Cincinnati Milacron, Coat-A-Matic and Trallfa. The list reflects the confidence of the manufacturers that the potential is enormous.

Case example of arc welding robot

Not all of the applications for robot welders are in the automotive industry. In fact arc welding is pretty ubiquitous in industry — there are 850,000 workers wielding arc-welding torches just in the USA.

At AiResearch, a division of the Garrett Corporation, a variety of fabricating operations are performed in connection with the aerospace industry. The company first introduced a robot into its plant not to reduce the labor content as such, but because of a chronic labor shortage in the area. The company is strong on welding experience, going back over 35 years, but they encountered a problem when they needed to weld mild-steel plates to hold electric motor stator laminations, used in rail vehicles. They installed a Unimate to make multi-pass fillet welds and actually had the robot alternating between two different types of motor housings with switched programs.

Welds which formerly took 5 to 6 hours to complete are finished by the robot in just over 2 hours, whereas another job which was taking 2 to 2½ hours is done by the robot in less than one hour. However AiResearch believes in getting the most out of its willing robot worker since periodically the welding torch is taken out of its hand and a routing tool placed in it, after which, by selecting the appropriate program, the robot is put to work cutting out a circular hole in an aluminium plenum chamber. Currently the robot spends 90% of its time welding and about 10% cutting.

Some technical points relating to the AiResearch installation may serve as a guide to robot capabilities. The motor housings are not small — they weigh between 750 and 900 pounds. They are brought to the welder by a jib boom hoist, and mounted on a positioner so that the necessary positional accuracy is achieved for the weld to take place. The welding position is horizontal, or at 45 degrees to that position. The robot provides signals which command the positioners to orient the housings appropriately. The thickness of metal welded is one half inch, and the material is low-carbon steel. However, other jobs require one inch steel to be welded. Not all of the welds are straight line. In fact one of the contours has two half-circle sections which welders found difficult to weld since the attitude of the gun must change constantly as it progresses along the path. The speed along the path ranges from 10 to 20 inches per minute and a power supply of 30 volts at 300 amperes is adequate for the job. The company states that the weld quality achieved is high, containing none of the irregularities which would be expected in a manual weld of this length. X-ray examination confirms the quality of the job performed.

Flame cutting: a related application

Iron and steel plates may be cut by the process of directing a closely regulated jet of oxygen on to an area previously heated to a cherry-red color, this representing a temperature above the ignition point for the oxygen. As the iron becomes oxidized the flame is moved uniformly to form a narrow cut in the material. Since only the metal in the direct path of the oxygen is acted upon, very accurate results can be achieved in cutting along a line if close control is exercised.

The analogy with welding is obvious. Consequently a robot can perform flame cutting to the same degree of accuracy that it can weld, and this is a field likely to grow. Programming will be no different from welding.

Chapter 13

Investment casting applications

The process of investment casting has its origins in the 'lost-wax' technique first used over five thousand years ago. Because of its ability to produce detailed and often delicate castings, the process was originally used in jewellery manufacture and for objets d'art, but its use faded out so that it remained dormant for hundreds of years until it was taken up again in modern form in the 1940's. Now it is in general use throughout manufacturing as a speciality process for making intricate parts of surpassing accuracy.

The investment casting process

The steps in the investment casting process are:

1 The production of an expendable pattern from a die
2 Coating or 'investing' the pattern with a refractory material to form a mold
3 Removal of the pattern from the mold
4 Firing the refractory material of the mold
5 Casting the metal.

The pattern material is usually wax, although sometimes this is mixed with a filler material to provide bulk and strength. The pattern is usually made by injecting wax under pressure into a machined metal die. A well-made die is capable of producing up to 50,000 wax patterns without being re-worked provided that it is properly maintained. As with die casting, die lubrication is important. For short runs and for prototype items where speed and low cost are important, adequate patterns can be produced from dies made from plaster, resins or low melting-point metals.

If high dimensional accuracy is required in the casting, then the pattern must be made to compensate for shrinkage of the wax and the metal being used as they both cool. Again, like die casting, entry ports for the molten metal plus vents to allow air to escape from the dies must be provided (runners and vents), while the same techniques are employed to make multiple molds which may embody dozens of identical small parts, joined by channels, all within the single mold.

When the part to be cast has been patterned in wax, the next step is to convert this pattern into a mold. This can be done by immersing the pattern in a slurry of fine refractory particles. After air-drying, or sometimes chemical hardening of the coat, successive coatings of slurry, often containing coarser particles of refractory materials, are applied to build up the thickness. The mold is allowed to dry out between each coat, and the result is to produce a stable, hard ceramic shell which totally encloses the wax pattern assembly. As its name implies, this is called a shell mold, though sometimes a more massive coating is applied so that the mold shows little external resemblance to the pattern. This is known as a 'block' mold.

Once the mold has been formed, the wax is removed by heating the unit in a furnace or by employing a pressurized steam method. In either case the wax melts and runs out of orifices provided. The mold is then finally fired at temperatures above 1000 degrees Centigrade.

Binder materials, usually silicates, are used to strengthen the mold, particularly with larger castings.

Finally the mold is filled with the molten metal selected for the casting. When this cools and sets, the mold is broken away. Why go to all this trouble to produce a casting? Firstly, the end-product requires little or no finishing. Extremely high accuracy of reproduction can be achieved with this process, and the casting bears no join marks which characterize a casting from a mold which splits into two halves. The surface finish and dimensional accuracy can be of an extremely high order. Finally, the initial die is machined only once, and then used to provide thousands of accurate patterns all of which are expendable. This contributes to lower die costs than would be the case if each casting required an accurately machined mold. The mold experiences hot wax rather than hot metals, so its life is greatly extended.

For quantity production, patterns are produced in clusters known as 'pattern trees', rather similar in concept to the multiple molds found in die casting. Then the trees are coated to form multiple molds prior to casting. The mold cluster will have a common entry port so that molten metal can be introduced at a single point to fill all of the individual sections. Figure 13.1 illustrates a form of multiple mold.

Preparation of shell molds takes experience, judgement and a critical eye. The work can be very tiring and boring, like so many industrial processes which are more and more being shunned by workers. Consequently the use of robots in this application is worthy of consideration.

After its long sleep, the investment casting industry is now very much awake and producing more than $250 million in cast parts annually. It is regarded as a growth industry, triggered off by the demands of the aerospace programs but now diversified in healthy fashion. Increasing labor costs have encouraged the growth of interest

Figure 13.1 *Example of multiple mold produced from pattern tree*

in the application of robot techniques.

Mold making by robot

When a man makes investment casting molds, the mold weight and silica coverage invariably fluctuate considerably, and this leads to the possibility that the castings will fail to meet the required specification. A scrapped casting, using this technique, means a lot of wasted labor since the mold is destroyed during the process and cannot be used again.

We have commented time and time again in this book on the fact that a robot cannot think for itself and, unless it malfunctions, cannot deviate from a program sequence which it has been taught and commanded to operate. This repeatability lends itself absolutely to the investment casting process for very obvious reasons. Once the robot has been taught the sequence of dipping, and then swirling each tree-like cluster of patterns, it will go on carrying out the same sequence without variation. This can only result in uniform molds, leading to castings which are consistent in their quality.

There are several functions which the robot can perform in this application, but they all involve transfers of parts between stations — a typical pick and place operation. All of the steps in the process which can be robotized are very labor-intensive, so there is good economic justification for the use of robots in this industry.

One of the most advanced foundry operations in the USA is that of a well-known marine engine manufacturer. Such engines rely heavily on cast alloy parts, and in order to improve its cost effectiveness, the foundry has introduced robots for making the shell molds. One of the features of the robot, and one which continues to crop up, is the ability to change parts simply by inserting a new program into the robot memory system. Since the robot is able to grasp any suitable part – in this case a part of the shell mold – only minor changes in program are generally required since the part can be designed to facilitate an easy pick-up by the robot. In practice at this foundry, program changes are typically made on-line by the foundry staff themselves. These simple program changes are not limited to purely positional functions, either. To take account of different shapes and weights of castings, foundry staff can very easily adjust the spin-time (rotational speed) of the pattern tree, this being part of the coating procedure, and they can set the depth of the dip into the slurry to ensure that the mixture does not find its way into the mold via the entry port (sprue) and contaminate it.

Add to this the fact that the robot can work hour after hour handling pattern trees which are too heavy for a manual worker to lift and the compatibility of the robot with the environment becomes obvious.

At the foundry in this example, the castings range in weight from about 4 ounces to 8½ pounds. Each shell mold requires six coats of slurry of varying consistencies and composition. The wax pattern is ultimately melted out in a steam autoclave, and the mold is then charged with alloys of steel, aluminum or bronze according to the nature of the part.

The present robot installation at the foundry uses robots to make the molds using pattern trees supplied by workmen. Racks containing as many as 30 trees are wheeled into the range of the robot and hand-loaded on to a pick up stand. When ready to coat the next tree, the robot reaches for it, and when through with the coating process, returns it to a similar stand with a positional accuracy of ± 0.05 inches. Initially this stand will be unloaded manually. After the final coat of slurry has dried, the wax is melted out and the silica coating fired at 1700 degrees Centigrade.

The robot controls the slurry mixer motors, the air valves used to fluidize the stucco bed and the gate valves in the fluidized bed dust control system. A special spin-control hand has been developed to hold and 'twirl' the shell mold (Figure 13.2).

Figure 13.2 *Special spin-control hand for 'twirling' investment casting molds*

In a later stage of automation a conveyor will carry the trees to the robot's pickup stand, and a second conveyor to take them away when coated. The conveyors will be monitored and controlled by the robot itself using external sensors and command signals generated from the robot memory. One nice feature will be that the robot will sense whether the place where it intends to drop a mold is in fact occupied. If not, it will move to an alternative spot as programmed into its memory.

Basic programs for robot mold making

The following programs specify the various steps carried out by the robot in conjunction with other equipment in the manufacture of high quality molds.

Program 1

FIRST COATING
Pick up tree from stand
Turn off slurry mix motor
Dip tree in fine grain slurry
Raise tree to allow slurry to flow evenly over surface
Restart slurry motor
Lower tree and spin out excess slurry
Raise tree to obtain even surface flow with remaining slurry
Move tree to fluidized bed of stucco (fine grain)
Turn on air to fluidized bed
Insert and withdraw tree from bed
Turn off air
Pause while excess stucco drains from tree
Place tree on stand
End of program

Program 2

SECOND COATING
Pick up tree from stand
Pre-dip tree into binder material
Raise and spin tree
Turn off slurry mix motor
Dip tree in fine grain slurry
Raise tree to allow slurry to flow evenly over surface
Restart slurry motor
Move tree to fluidized bed of stucco (fine grain)
Turn on air to fluidize bed

Insert and withdraw tree from bed
Turn off air
Pause while excess stucco drains from bed
Place tree on stand
End of program

Program 3

THIRD, FOURTH AND FIFTH COATINGS
Pick up tree from stand
Turn off slurry mix motor
Dip tree in coarse grain slurry
Raise tree to allow slurry to flow evenly over surface
Restart slurry mix motor
Move tree to fluidized bed of stucco (coarse grain)
Turn on air to fluidize bed
Insert and withdraw tree from bed
Turn off air
Pause while excess stucco drains from tree
Place tree on stand

Then, for the sixth coating, the first part of program 3 is repeated but the steps used in applying the stucco coating are omitted.

Case example at Pratt & Whitney

In the process of making shell molds, the robot is the central figure. The ancillary equipment is relatively inexpensive and prosaic. The robot is choreographed in an elegant dance that is mated to the job in hand. Quality improvement is so dramatic that even a single robot in a plant can be justified.

However, as an example of a company making full use of the robot's abilities in this application, and as an indication of both the growth of importance of the investment casting industry and the application of robots, recent developments with the Pratt & Whitney Aircraft Group are of great interest.

The company needs to produce exceptionally strong aircraft turbine blades in very large quantities. To this end they designed not only a new form of blade, but a whole new process for manufacturing it — the investment casting method.

These blades have an advanced metallurgical structure based on single crystal configurations, hitherto the province of the semi-conductor industry. The design permits the blade to be made in two halves, and the manufacturing complex is entirely automated. After each half of the blade is duplicated in wax, the pattern thus formed enters a line of no less than ten robots where in about six hours, while each piece is passed from robot to robot, a layer of ceramic slurry is deposited on the patterns to a depth of about one quarter inch. The final operation is to coat each pattern with fine sand and then dry it, after which the wax is removed so that the mold so formed can be used to cast the blade section.

An incredible production rate of 50,000 blades per year has already been achieved by the present facility, but it is intended to increase this to almost double that capacity, sure evidence of the success of the robot line in this application.

Chapter 14

Forging applications

In forging, metal objects are shaped by hammer blows or presses. The metal to be forged is heated to a temperature at which it is in a plastic condition, and therefore able to be shaped by hammering.

The technique is among the oldest known, the blacksmith having for centuries tended his furnaces to heat iron before beating it into shape on an anvil. The modern counterpart of the blacksmith's fire and bellows in industrial forging are large efficient furnaces, while his hammers have been mechanized. Machines capable of exerting rapid hammer blows and many tons of pressure are commonplace in the forging industry today.

An advantage of forging is that the grain pattern in the metal can be re-arranged by the heating and hammering treatment so that the strength of the part formed can be optimized along various axes to meet the stresses which it will encounter in use. As a result, components produced by forging can be much stronger than they would be if they were machined from cold stock.

Forging processes

Forging is no longer a single process. Through mechanization, it has developed into several categories, namely:

1 Drop forging:
    (a) Hammer forging
    (b) Die forging
2 Press forging
3 Upset forging
4 Roll forging

A few words describing each of these processes will be useful in assessing the role of robots in the forging industry generally.

Drop forging

HAMMER FORGING
The hammer forging method is the one most resembling the work of

the blacksmith. It is very suitable for producing simpler shapes, often those which will be machined to final dimensions after the forged part is produced.

A drop-hammer is used, the lower part of which is the anvil while the hammer or ram is raised and dropped between vertical guides.

The heated billet of metal being worked is struck by the hammer, and simple open dies, often no more than rectangular guides, may be utilized to produce the shape required. Billets are forged into rough shapes by repeated hammer blows during which the piece is moved back and forth and turned from side to side by the operator who judges the number and strength of the blows struck and the position of the billet with respect to the hammer.

Drop hammers are powered in various ways such as by steam, air or mechanical systems. When steam or air are available, they can be used to form a cushion to mitigate and control the force of the blow. Smaller machines often use springs for the same purpose. Since the hammer may weigh anything up to 50 tons in large drop forging installations, the impact this produces demands that the machine be of massive construction set in very solid foundations.

DIE FORGING

In die-forging the technique is much the same except that the hot metal is more accurately shaped by being hammered into two-section dies which eventually must close over the billet. The bottom section of the die rests on the anvil and contains the billet; the upper half forms part of the hammer.

Die-forging clearly calls for greater precision if the die halves are to close accurately on each other. Adjustment must be provided to ensure that the die registration can be maintained as the machine and the dies are subjected to wear and tear.

Dies obviously take a beating. They are usually made from high-grade steel, heat-treated, ground and polished. Within the die there may be separate areas through which the part progresses as it changes shape from initial billet to final forged form. The operator must move the part from place to place between hammer blows, and more often than not this has to be done at a rapid pace. Some excess metal or flash is inevitably produced, so the die must be designed to permit this to be trimmed off while the part is still hot.

The operator (or hammersmith) works the drop hammer by means of push buttons or a foot-switch. He controls not only the number of blows but also their force, and his job is a skilled one. The fewer the blows needed to shape the part the higher the production rate and the less the wear and tear on the dies. Even so, the hammer may make as many as 300 blows per minute and achieve pressures up to 50,000 pounds per square inch.

Some idea of the versatility of the process may be gained from the range of parts which are able to be forged. Pieces weighing a few ounces up to several tons are regularly produced by drop forging, though hardly all in the same size of machine.

Press forging

In this type of forging the hot metal is subjected to a relatively slow squeeze exerted by a hydraulic press. Practically all of the energy from the press is absorbed in the workpiece, unlike the drop hammer method which distributes the force of the blow between billet, machine and its foundations.

Presses delivering up to 25 tons per square inch are possible, and the method can be used to make really large parts requiring ingots weighing 200 tons or more. Typical examples are naval guns and marine engine crankshafts.

Press forging should not be confused with press operations on sheet metal such as are used in the manufacture of automobiles, aircraft and domestic appliances. Such pressings are carried out on cold metal which is deformed to the required shape by (usually) a single sharp closure of the press jaws.

Upset forging

This is a somewhat specialized form of forging used mainly in the nut and bolt industry. It is limited to the forming of parts which have a shaft terminating in a larger head. A length of circular or other cross-section rod is gripped by two dies of selected shape. The dies open up at one end, and at this end a ram or header is driven several times to 'upset' the end of the rod which is forced into the shape governed by the large end of the die.

The process is very easily automated and header machines are used in large numbers in the fastener industry. The part to be made can be changed simply by changing the dies so that machines are suited to short runs.

Roll forging

This is another specialized process. As its name suggests it is a technique for shaping hot metal by rolling it. The rollers can be machined to give the required shape, but the method is obviously limited in the number of shapes which it can produce just as a lathe is limited to parts of circular cross section.

The working environment of the forging process

The hot forging of metal represents one of the most hazardous of all industrial environments. It is noisy, hot and the air is polluted. The work is extremely repetitive which makes it unattractive to new generations of workers, and despite the advent of mechanical aids of various types, it demands a certain amount of physical strength to manipulate both the machinery and the forgings in many shops. For these reasons the forging industries have long recognized the need for greater automation, but this is impeded by the fact that many of the products are short runs.

Forging would therefore appear to be the ideal application for an industrial robot if it were capable of enduring the debilitating environment.

However, there are several conditions with which robots cannot cope but which can be tolerated by human operators. In open die forging we have seen that there is a need for the operator to move the casting about and to vary the hammer blows in order to forge the shape needed. In this application the operator is working exactly like the old time blacksmith except that mechanization has removed the need for the hammer to be swung either by the smith or his mate. This ability to see the part and make an instant judgement on the strength and place of the next blow is beyond the robot's capability, and outside the scope of automation systems too. Speed is another aspect to consider. Human operators can move with amazing alacrity when necessary, especially when the forging of small parts is being done. This is often necessary simply to complete the forging before the part cools, and this type of operation is impossible for the robot. Robots capable of lifting a heavy part at the end of a long arm tend not to be fast movers.

Possibly the role of the robot in this industry is simply to support the human operator, and this is already being done, not by robots but by manipulators which are under human control.

To complete the list of problem areas in forging, consider the following points, each of which makes the going difficult for the robot. First, much forging equipment is archaic and many of the dies in use are marginal. As a result, the part often sticks and does not eject from the drop hammer in predictable fashion, staying in either the upper or lower die. Secondly, in closed die forging, a part may be made by several sequential operations during which the shape of it may change dramatically. This may call for one or more completely new sets of grippers if the robot is to be able to get hold of it at all stations down the line. Thirdly, billets are often heated in gas-fired furnaces which are loaded manually. The billets are stacked in almost random fashion so that the robot would not know where to grasp the part. This can, however, be avoided by the use of induction heating, the parts being laid out in more orderly fashion.

All of this seems to sound the death knell for robots in the forging business, but happily this is not the case.

Robots in forging

Despite the fact that forging presents problems for robots, they are already working in this industry and it is certain that their numbers will grow and the complexity of their operations will continue to increase.

The robot can pick up hot metal parts and can work for long periods in hostile environments. These characteristics have been put to good use in a company which forges high pressure gas cylinders on a production basis. It freed two operators to be assigned to more acceptable jobs and contributed to a labor saving which should result in a payback of the robot's cost on the basis of labor-savings alone.

In this application the human operators had to use a hoist to transfer hot metal preforms from a 1500-ton forge press to a 260 by 100-ton draw bench where each preform, weighing between 60 and 360 pounds, was drawn into cylindrical shape of required size. Now, a robot receives a signal from the forge press to tell it that a preform is at the required temperature. A piston pushes up the preform so that it can be grasped by the robot. A typical preform weighs 143 pounds, and the robot picks it up, swivels through 180 degrees, places it on the draw table, withdraws, swivels back, immerses its gripper in a tank of cooling water and then waits by the forge press for the next signal. In this application, the preform stays in the forge press for 45 seconds, while the robot cycle takes only 10 seconds. There is about 35 seconds of wasted time as a result, but even with this disadvantage, this is one application of a robot in forging which is making economic sense.

In another application which is only marginally a forging one, the robot again uses its ability to handle hot metals. In the manufacture of a chain, a rod is heated to about 1600 degrees Fahrenheit and ejected on to a mandrel where a robot picks it up and takes it to a welding position. When the rod is welded into a closed ring, the robot takes it to a point where an open ring rests on a closing machine. The robot places the ring over the open link which then is closed by a signal instructing the machine to operate. The robot then returns the parts to the welder where the open ring is welded. The links are, in addition, turned over 180 degrees by the robot to permit welding on all sides. The robot continues in this fashion, and as the length of chain builds up it is supported to avoid overloading the robot. The robot deals with links of 3 to 5 inches inside diameter. Figures 14.1 and 14.2 illustrate this application in plant layout and the special gripper needed. The robot depicted in this case is a Unimate 2000B.

In yet another application, an automobile manufacturer is using

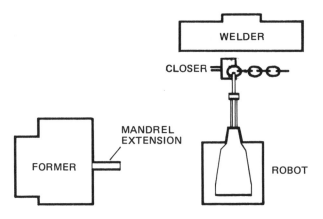

Figure 14.1 *Plant layout for robotized chain link manufacture*

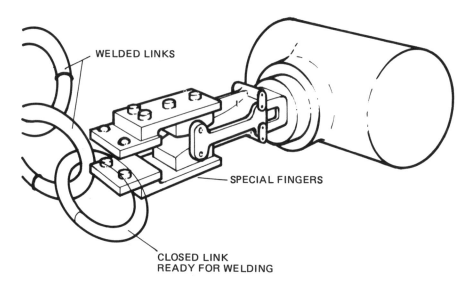

Figure 14.2 *Special gripper for chain link manufacture*

a robot in a high production, hot forging operation — in die forging in fact. The robot picks up a slug of hot metal — at 1800 degrees Fahrenheit — and positions it in the forging die. It receives two blows, then the robot transfers the part to its next location. An operator is still used to monitor the critical heating process, but no longer has to place his hands in or near the die of the forging press, a potentially very dangerous place. This installation turns out 12,000 parts a day on a 24 hour basis.

A more comprehensive use of the robot in a similar application is to be found in an Italian automobile plant. Here a forging press is used rather than a drop hammer. A vibratory feeder delivers billets to an induction furnace which heats them to 1200 degrees Centigrade. From here a feeder system presents them to the robot in a simple yet effective coded form; according to the way in which the billets are to be forged, they are presented either horizontally or vertically.

The first operation is to shape a square billet into cylindrical form in three stages, blocking, intermediate forging and finishing, all press operations. As the billets emerge, the robot waits for them and places them vertically in the first die, withdraws, and then generates a signal telling the press to close. When the first operation is complete, the opening of the press generates a signal which instructs the robot to take the billet and place it in the second die. As it withdraws, it again signals the press to close. When this operation is complete the same sequence is employed to transfer the part to the finishing die. On removal from this die, the robot places the part on a discharge chute which conveys it to a trimmer. Future development of this process will be to have a robot carry out the trimming operation and finally place the part in a container.

In another forging shop a robot is carrying out a similar series of operations but in a drop hammer machine rather than a press. The robot positions the part in appropriate dies, and waits while the programmed number of blows have been struck. The difference in this case is that the furnace used has been specially developed for this type of fast operation and contains a magazine of billets sufficient for one hour's operation. Another innovation is the inclusion of a photo cell device in the transfer line to sense the arrival of a hot billet and initiate a signal to start the forging sequence. Typical grippers for this application are shown in Figures 14.3 and 14.4.

To sum up, robots are good at handling hot metal parts, are capable of accurately positioning those parts and of responding to signals from other equipment used in a multi-machine installation. They are unaffected, to a large degree, by hot, dirty environments and polluted atmospheres, though not entirely so and special design characteristics may be required if the environment is too hostile. What the robot lacks is a real turn of speed such as a human operator can show when needed, but on the other hand the robot does not tire easily and far outpaces the human operator in its endurance. However, the robot is also blind, and is consequently not able to make visual judgements as are frequently used by operators in determining precisely how to forge a part.

The forging industry needs robots. Older, experienced forge operators are retiring and not being replaced. Techniques are improving and new metals being forged so that the industry is far from played out. Indeed it represents a challenge to the robot designer because of its technical problems and its potential market.

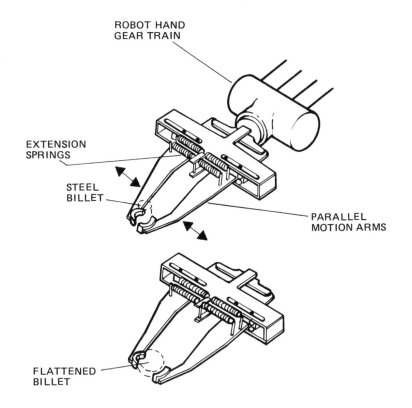

ROBOT HAND
GEAR TRAIN

EXTENSION
SPRINGS

STEEL
BILLET

PARALLEL
MOTION ARMS

FLATTENED
BILLET

Figure 14.3 *Special hand design for holding hot metal billets*

Figure 14.4 *Robot handling cylinders in forging operation*

Chapter 15

Press work applications

Presses used industrially to form and shape metal consist of a solid bed on to which a rigid movable platen closes, usually under hydraulic power. Dies of appropriate type are inserted between the two sides of the metal to achieve the necessary shape. These dies may actually cut the metal (stamping and punching) or deform it by stretching the material and producing the required three-dimensional shape.

Press operations

The method is capable of high accuracy and speed. A part forged by drop hammer (Chapter 14) will sometimes be inserted into a press for final finishing. Seldom is any hand-finishing needed after the work is removed from a press. Much press work is completed at a single stroke (single action press) but two or even three press operations may be required according to the strength and complexity of the part. Some presses use rubber pads attached to the upper (movable) platen. Contrary to popular belief, rubber is practically incompressible but it will change its shape readily under pressure and flow in all directions, transmitting pressure as it does so, much in the same way as a hydraulic fluid.

Various presses are available to meet the wide demands of industry. They range from giant 5000-ton models exerting about 2000 pounds pressure per square inch on to the work down to small, single-action units which work at very fast rates.

The whole process is reminiscent of forging. A press resembles a slow-moving drop hammer. Press forging is obviously directly analogous. However the major difference is that in press work the workpiece is not heated but shaped or cut at room temperature. It is a matter of judgement and experience just how far a given metal can be deformed before it will tear.

Press operation is a vital process in the manufacture of automobile body panels and other car parts, aircraft structures and many domestic appliances. When a large press is available it can often be used economically by making small parts en masse using multiple dies in just the same way that this is done in die casting (Chapter 10) and investment casting (Chapter 13).

Some press operations involve passing the part sequentially through three or more separate presses. Whatever the operation, the method used is essentially the same — pick up the stock (usually flat sheet), place it between the press faces in correct registration with the dies, operate the press, remove the part, stack it or, if the operation is a sequential one, place it in the next press. In robot parlance, these are 'pick and place' operations.

Apparent opportunities for robots in press work

Press operation is a dangerous job. Sheet metal stock often has sharp edges which can cut operators handling it. The press, once operated (through a clutch mechanism which transmits the drive) is remorseless and will sever the limbs of any operator unlucky enough to be caught between the platens as the press faces close. Strict safety regulations govern the use of such presses, although unfortunately these are not always enforced. Typical among the requirements for safety are electronic devices which make the press inoperative when hands or other parts of the operator intrude during a press cycle. The risks would be thought to be sufficient to justify the use of robots in press work regardless of cost, but this has not been the case to date.

In the early days of industrial robots it seemed quite logical to assume that they would be in great demand for press loading and unloading, but this was to some extent an unfulfilled aim. The great preponderance of press operations requires that the pieces of stock be fed manually because they are not orientated accurately enough at the input station for the robot to be able to grasp them reliably. Much of the operation is single shot, which leaves little opportunity for the robot to engage in press-to-press transfer where it would be expected to come into its own. Another consideration is the speed at which many presses operate, much too fast for a robot. However the robot has found its way into the press shops and is particularly at home where the parts are larger and the cycle time correspondingly slower. Its main role is in loading the press, although load-unload and press-to-press transfers are not uncommon. It is all a question of economics which is in turn dependent upon production rate (or press speed). If, for example, the robot has to go through a complicated maneuver to turn the part over in moving it from one press to another, this can be time-consuming in an industry where a five-second transfer time is not regarded as particularly short.

Finally, the robot, as always, must compete with standardized automation equipment in the form of stacking machinery and limited-sequence machines, all of which are used extensively in the industry. Despite the apparently gloomy outlook for robots, they have found applications in press work and their infiltration will undoubtedly continue to increase. We can describe a few typical jobs which they are carrying out successfully.

Current applications of robots in the press shop

Once an application for a robot has been identified in a particular press shop — and as we have seen this will depend on various factors such as speed and part-orientation — then one can look forward to consistent operation, shift after shift with none of the problems of employee safety which attend a manually operated press line. Unlike human operators, the robots will not lose their concentration through fatigue after a busy shift. Nor will they attempt to 'cut corners' but instead go steadily through their programmed movements hour after hour. Changing to a new program takes only minutes.

Robots in press shop at Ford Motor Company

In the Ford Motor Company, presses that once required four-man crews can now be operated by two men, thanks to the help of robots. In this case, heavy rear-suspension parts are produced on two presses, each linked by a conveyor. Each press has two die-stations. The parts are placed into the first die station in both presses by manual operators. The presses then operate and the pieces are ejected automatically. At this stage the robots pick up the parts, turn them and position them with high accuracy in the second die station.

The action of the robots is synchronized with the operation of the press rams through limit switches which prevent the robot arms from entering the die areas until the rams are located in the up position. Similarly, limit switches sense whether parts have been loaded into the second die stations to avoid double stacking which would result if a robot operated before the die had been cleared.

The whole operation is a classical example of a gradual introduction of robots into an established industrial process which in this case was already partially automated. The results are reduced manpower and a greater safety factor, but productivity has also increased. It had been found that when human operators loaded parts into the second die station they were not always positioned accurately enough at the first attempt, and hook-like tools were used to manipulate the part into the correct position, a very time-consuming process. The robot gets it right almost every time so that the hourly production rate increases. The application shows how men, robots and automation systems can be effectively combined with better safety, increased production and a resultant improvement in the working environment all at an economic cost, but this is only the beginning. Figure 15.1 shows the two robots in

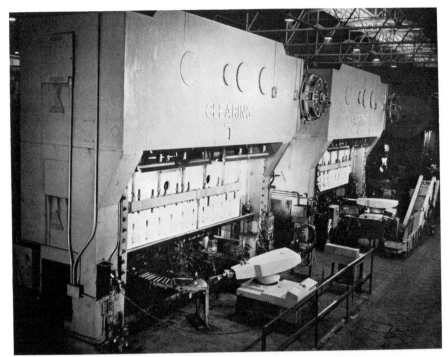

Figure 15.1 *Robots in the Ford Motor Company press shop at Dearborn, Michigan*

Figure 15.2 *Robot in press shop fitted with two hands*

the installation described, almost dwarfed by the giant presses they serve. In Figure 15.2 a Unimate robot fitted with two hands to provide high-speed transfer of smaller parts through three presses is pictured.

Robots installed to meet safety regulations

In another application, a manufacturer making very heavy parts using 800 ton presses decided to instal robots in order to meet the safety regulations. Senior executives in the company felt this to be an expensive solution to the safety problem and were by no means convinced that the decision was a good one. Six robots were purchased; of these, two were positioned on each of two press lines and another operated between the two lines. The sixth was a back-up unit which was intended to be substituted for any robot in the line which had to be pulled out for maintenance. The proposed cycle times were not dramatic — about 340 to 360 parts per hour to be produced with the robots carrying out a press load and unload function. This production rate was fully 100 units less than could be achieved using human operators which accounted for the lack of enthusiasm on the part of certain of the management team when the robots were installed.

The parts themselves were automotive components weighing about 24 pounds, a weight well within the lifting capability of the robots used, even when they were lifting them with fully extended arms. The robots chosen by the company had six programmable degrees of freedom. It should be noted as a general rule that if a robot is to be located off the center line between the beds of two presses, then five articulations are almost always necessary to gain proper orientation of the stock in cases involving press-to-press transfers. If a turnover of the part is involved six articulations may well be required. Since this fact is not always recognized by potential buyers, the decision by the manufacturer in the application described to instal six-articulation capability showed considerable knowledge and foresight. Another factor influencing the decision was the aim of providing maximum flexibility in the robot within the acceptable price-bracket so that it could adapt to future changes in the press-line layout which might be necessary.

The company using the robots in this application was already using numerically controlled machines extensively and had a comprehensive in-house maintenance schedule into which the robots could be fitted without much difficulty. Now, after each robot has completed more than 20,000 hours of operation (about half the projected life of the machine) maintenance costs have worked out at about $3000 per unit per annum, rather less than $4.00 per hour of operation.

As to results, the slower production rate has been to a very large extent offset by the reliable steady output of the robots, though there have been some problems in getting the robot to pick up and place accurately these heavy parts. However each robot replaced two men per shift, or six per day when working three shifts. With five robots operating, this means a saving of 30 operators per day, while the only support needed by the robots was one maintenance man on each of the first two shifts, both of whom can do other work if the robot does not require their services. It seems clear that these savings will more than justify the purchase price of the robot.

The sequence followed by the robots in this installation, two on each line, is as follows:

ROBOT NO. 1

1 Interrogates first step to see if interlocks have signalled that the ram of the press is on top center in the range 300 through zero to 5 degrees. Also checks that the press made a complete rotation indicating that a part had been formed.

2 If all requirements of Step 1 are met, robot advances into the press. As it does so it activates a circuit which immobilizes the clutch mechanism, making any running of the press impossible.

3 When in position, robot grasps the part and resets the conditions of Step 1.

4 Robot raises the part.

5 Robot retracts the part and interrupts a photocell beam which is a signal that it has the part in hand and the next step can therefore proceed.

6 Robot boom rotates through 180 degrees and the wrist rotates the part 180 degrees. As these two motions are simultaneously being completed, the robot resets the circuit which allows the press to cycle once again. It also scans the second press to ascertain that the ram is at top center and if the part has been removed from the die area.

   When satisfied that all these conditions are met the robot then moves to the second press.

7 As robot moves to the second press it immobilizes the clutch to prevent the press from operating.

8 When in the correct position, it sets the part down.

9 Robot releases the part from its grasp.

10 Robot retracts from the press.

11 Robot rotates 180 degrees, taking it back to the first press. In so doing it also activates the clutch mechanism for the second press so that the ram comes down as the press goes through its cycle, after which all conditions met in Step 6 are re-set. This completes a full cycle of operations for Robot no.1.

ROBOT NO. 2

1 Robot interrogates first step in its program to see if interlocks have signalled that the ram is at top center and whether it has made a 360 degree rotation indicating that a part has been formed.

2 If the conditions of Step 1 are met the robot advances into the press. As it does so it activates a circuit which immobilizes the clutch mechanism, making any running of the press impossible.

3 When in position the robot grasps the part and resets the conditions of Step 1.

4 Robot raises the part.

5 The robot retracts and interrupts a photocell beam which is a signal that it has a part in hand and informs Robot no.1 that the part has been removed from the die area.

6 Robot boom rotates through 180 degrees and at the same time resets a circuit which allows the press to cycle once again. Robot no.2 then looks at the third press to note whether the ram is at top center and whether the last part has been removed.
   When satisfied that all these conditions are met Robot no.2 then moves to the third press.

7 While moving to the third press it immobilizes the clutch to prevent that press from operating.

8 When in the correct position, it sets the part down.

9 Robot no.2 releases the part from its grasp.

10 Robot retracts from the press.

11 Robot rotates 180 degrees back to the second press and in so doing activates the clutch mechanism for the third press so that the ram comes down as the press goes through its cycle, after which all conditions met in Step 6 are reset.
   This completes the full cycle of operations for Robot no.1.

The above two applications show that the robot is beginning to find a place in the press shops of industry, but much work remains to be done before robots fully meet the requirements of the press process. Even the leanest, fastest robot is a sluggard compared with a human operator who can feed a press making 12 strokes a minute — that is one every five seconds! But there are other issues which militate against the robot having an easy entry into the field.

Outlook for further robot handling of press work

In a press transfer line the first station is probably best handled by stacking equipment designed to deal with flat stock. Such equipment exists and is not robotized. There are hard automation concepts deeply entrenched in the industry which robots will find difficult to oust; indeed they should not be forced out until the robot can do a better job — which includes higher productivity and larger profits.

This is especially true of situations which involve the sequential passing of parts through three or more stamping presses; systems built by Danly are particularly suited to this operation. However the cost of a hard automation line has continued to rise, whereas the robot designs are resulting in mass-produced models which make them competitive pricewise especially when one takes into account their ability to be reprogrammed instead of being consigned to scrap. Against this apparent cost advantage however one must realize that very large press lines are often installed on a turnkey basis and the massive investment in the presses themselves tends to swamp the cost of the automation equipment.

Robot handling of larger parts is feasible, but when it comes to small parts with very short cycle times the 'heavies' are ruled out. Some of the business has gone to limited sequence robots. In Japan, the AIDA company which manufactures stamping presses also supplies pneumatically operated arms which can be used to feed die to die in a single press and in turn be manually fed themselves. Their true role is safety, making it unnecessary for an operator to put his arms into the die area.

Volkswagen is working on a unique form of construction in which a single arm is mounted at its center and is free to extend in two opposite directions with grippers attached to each end of the arm. The configuration has the advantage of eliminating the round trip between presses which characterized robots of the one-armed type such as Unimate. So far the concept, though sound, appears too slow.

In an attempt to solve the same type of problem, Kawasaki is working with a mechanical configuration which may be more appropriate to press working than the polar coordinate configuration in robots such as Unimate. In this instance, there are two arms located between presses and an intermediate parking station between the arms. Thus parts move from press to press in a manner quite similar to the motion of a Danly automation line, but the arms are reprogrammable so that the press transfer line can be set up for different parts as dies can be readily changed. The intermediate parking station also eases the problem of part turnover. It is too early to say whether the approach adopted will be cost effective.

Other robots aiming for the press market are Cincinnati, PRAB and Versatran, and Unimate has already been mentioned as one which has achieved a measure of success in the application.

The Unimate kinematic construction coupled with its high-speed in and out motion is probably superior to the revolute construction of the Cincinnati robots. PRAB is built on similar lines to Unimate from a kinematic standpoint, but is much slower. Versatran, working in a cylindrical coordinate concept, could be well suited to the application except that to date it has lacked the reach needed for most press work.

Press transfer is a challenging field for the robot designer and not at all the simple job requiring only limited manipulative ability which many people assume it to be.

Chapter 16

Spray painting applications

Painting, so the dictionary tells us, is the coating of surfaces with a liquid mixture, usually of a solid pigment in a liquid medium, for the purposes of decoration or protection. The description is apt and will be immediately recognizable. The process goes back thousands of years, certainly to Egyptian times as the relics of those dynasties prove.

Up to the early nineteen thirties, paint was generally applied by a brush, and it took some hours to dry. This created problems for the expanding automobile industry since car bodies, after painting by hand, had to be left until the paint hardened. This required much non-productive floor space and represented a critical factor in production rates. Moreover the painted surfaces of the day were prone to collect dust, or to run, so that the chances of a body being spoiled were considerable. It is not surprising, therefore, that much research and development was devoted to the technique of painted finishes, with the result that quick-drying paints applied through the medium of a finely divided air/paint spray mixture became the accepted method. Today spray painting even in the home environment is commonplace through the use of aerosols or simple home workshop types of compressor and spray gun.

Paint behavior and the technique of painting

Almost everyone has at some time or another attempted some form of spray painting. It soon becomes obvious that to avoid runs and to get an even surface the spray gun must be kept on the move, and a great deal of skill is needed to obtain a 'professional' looking finish. It is perhaps the automobile industry which has set the standards. It is unthinkable that one would accept a new car which did not have a superb painted finish; even an old car, sent for repairs, is expected to be returned in pristine condition. The technique is not an easy one, but it has been helped greatly by the development of suitable paints which flow easily, minimize runs and dry rapidly before dust can settle on the wet surfaces. Nevertheless the painting of car bodies is still a major problem in the industry especially now that production rates are expected to be so high that only a small amount of time is available for the paintwork to be applied and completed, in not one but several coats.

Spray painting is not, of course, confined to the automotive industry but is to be found wherever the provision of a high quality painted finish is required on a mass-produced item.

Nor are paints the only finishes to be applied by spray techniques. Enamelling and powdering of surfaces can be dealt with in the same way, and a typical example is the application of vitreous enamel to domestic bathtubs. It is important to realize that the movement of the spray gun as it is controlled by the operator is a three-dimensional process; coating an object fully, especially if it has complex contours, requires movements in depth as well as laterally. Also the use of the wrist is significant in that it is often necessary for the operator to angle the gun in a particular manner to deposit paint on an inaccessible surface.

It is of interest to investigate how a paint-spray operator goes about his business. Holding a gun pressurized by air and fed by paint from either a small tank on the gun itself or a central reservoir, the painter depresses the trigger on the gun to release a fine spray of paint carried in the air stream. The viscosity of the paint must be just right to form a fine spray and not be ejected as lumps or globules to mar the surface finish. The distance from the work at which the spray is projected is also critical in order to produce a coating which is thick enough to cover the surface but not so heavy that unsightly runs occur. The operator can control the process by constantly moving the gun, applying thin rather than thick coats of paint, going back over areas already thinly coated to build up a layer of even thickness and good finish.

The path of the spray, though not random, is not precisely defined. If the path traced out by the tip of the gun were to be recorded it would be found that even the most experienced operator painting the same surface day after day would never make two identical passes over them. Instead he would use experience and judgement and by looking at his work, make instant decisions where to spray and for how long. In other words, the technique is more of an art than a science. This is a consideration when evaluating the use of a robot in the spray painting application.

The spray painting environment

The spray painting environment has always had the reputation of being one of the worst which human operators have to encounter. To maintain a dust-free painting area at the right temperature means that the 'shop' should be as restricted in size as possible. Some early paint-shops were real death-traps, but nowadays legislation has produced codes aimed at insuring the health and safety of the operator.

Solvents used in painting are toxic and produce a polluted atmosphere, so ventilation must be provided which gives an abundant supply of fresh, clean air to the operator. The optimum paint-shop layout to provide this may be not at all the best arrangement for production-line painting, however, so compromises are necessary. The wearing of masks by painters has been de rigeur since the inception of the technique.

More recently, attention has been paid to noise, and the noise-levels generated in a busy spray shop, arising from the air discharge through fine nozzles, can, in an eight-hour shift, cause irreversible damage to the ears. As a result, current standards require operators to wear ear-plugs. Yet another problem is fire hazard, arising from the highly flammable nature of the materials used.

Finally, just to high-light the particular unpleasantness of the environment, certain of the pigments used in the technique are suspected of being carcinogenic agents. This, together with proposed new regulations governing hydrocarbon emissions from manufacturing facilities, will probably result in major changes of practice in the industry, such as the use of waterbase and urethane materials. The whole field is therefore seen as one for further research and development.

Automation in the paint spraying industry

The hostility of the paintshop environment has always made it a prime candidate for the adoption of techniques which might take the human operator out of the process so that the implementation of many of the costly protective measures now demanded could be avoided.

Automobile production with its competitive approach has been a fertile area for such developments, and many spray painting machines have been introduced. Generally speaking, however, they are limited in their use so that they almost always have to be backed up by human operators who can touch up areas missed by the machine. The machines also tend to be more wasteful of paint; typically they are designed to paint with horizontal and vertical paths on a reciprocator system, that is a back and forth motion. Batteries of spray guns fed from large capacity centralized paint reservoirs move according to a pre-determined program and manage to paint some 70% to 80% of the exterior surface to be covered. Less accessible areas such as the wheel arches, inside the trunk and engine compartment and door edges, to mention only a few, must be painted by operators who look for unpainted areas as the car body leaves the automatic painter. Color changes have to be properly scheduled, since this involves changing the paint in the reservoir or switching to a new reservoir, and making sure that the old color has been cleaned out of the paint lines and guns to avoid contamination.

The method is well-established in the industry yet recognized as being far from perfect. The energy requirements are enormous. It takes about 25 million BTU's to build a car. Much of this demand comes from the need to supply fresh air to the spraying booth during finishing. Since the U.S. government will be demanding that this be cut by about 20% by the early 1980's, it is clear that there is a considerable incentive for the automobile manufacturer to get human operators out of the paintshop wherever possible, quite apart from the economic advantages.

There is little doubt that the robot offers the best chance for success for the car builder, faced as he is with constant design changes, new color and finish fads and a buyer who seeks high quality at low cost from a product which must look good as well as perform adequately.

Robots in paint spraying

In the many pick and place operations which robots are called upon to perform it is the starting and end points of the motion which are of importance. The path traversed to travel between the points is generally unimportant — though not always, as we have seen with seam welding (Chapter 12). For spray painting however the path is the key; in a sense the end points are almost unimportant since a paint trajectory could theoretically start and finish outside the confines of the workpiece.

Robots having such capability are the Continuous Path (C.P.) machines. Other types of machine can be programmed to apply paint from a spray, but the C.P. machine most nearly emulates the action of a human operator. In fact such a machine can be taught by having an expert painter lead the robot in its learning mode through an actual paint job, after which, unlike the human operator, the machine will endlessly duplicate the lesson, achieving a positional accuracy better than 2 mm. throughout the program.

A disadvantage of this type of robot is that if one part of the program should prove to be unsatisfactory, it cannot be changed without the entire program being rewritten, but in practice this is not a very serious matter since once the program is right, it can be expected to be utilized many thousands of times.

In the USA an average spray booth in the automative industry is of width 15 to 17 feet. This presents some restriction on the robot's movements, and tends to rule out some of the available larger machines. Redesigning the booth can be very costly, so the better solution is to design the robot to work in the available space. The outcome is that special robots have been developed for this application, the best-known among them being the Norwegian robot made by Trallfa Niles Underhaug and marketed in the USA by the DeVilbiss Company. The Trallfa robot completely dominates the paint spraying industry at the present time despite competition from certain European and Japanese companies. This comes about largely because the company has dedicated its product to this one application rather than attempt to provide a multi-purpose robot.

The Trallfa paint-spraying robot consists of two main units. The first is the mechanical-hydraulic unit with its single operating arm, as shown in Figure 16.1. The second unit is the electronic control including the magnetic tape memory. The active part has five movements, which in human terms are the horizontal hip, the shoulder, the elbow and two wrist movements, side to side and up and down, each of these traversing an arc of about 200 degrees.

Figure 16.1 *The Trallfa spray-painting robot*

Each of these joints associated with these movements is equipped with a resolver which measures angles by a phase method, actually the time difference between the phases of a reference voltage and the corresponding resolver voltage. This difference is fed into the electronic control unit and used to determine the direction and amplitude of the robot's movements.

The programming of the robot is done by an experienced paint sprayer who, with the robot on low power so that the arm is more or less weightless, leads the unit through movements necessary to give the workpiece one coat of paint. The robot is now ready for work and will repeat the program as taught on receipt of an appropriate command which can be, for example, a photo-cell signal indicating that a job to be painted has come into position on the production line.

Programs are changed by removing a tape cassette and replacing it by another so that, if necessary, many hundreds of different programs can be stored away as the robot's repertoire increases.

It is important to make the distinction that the Trallfa robot is not a spraying system but a robot which replaces the spray gun operator. It can use any type of gun and, within the limitations of its reach (just like a human operator) spray any surface with any form of protective coating.

While the Trallfa robot may be unaffected by polluted atmospheres, it has to operate in an environment where a spark could cause fire or explosion because of the solvents used. The robot has therefore been designed with the aim of minimizing such risks.

It is an interesting comment that just as some humans are singled out early in life to train as specialists in certain fields, so this robot was prepared for its main job — painting. There are other jobs which it could do, just as an accountant might make a reasonable attempt to lay bricks, but its *pièce-de-résistance* is its deft wielding of the spray gun.

Outlook for robot painting in the automotive industry

Whilst the use of robots in spray painting will undoubtedly increase generally, there is one industry — automobiles — where they are going to make a great impact in the future. This statement is based on the fact that robots have already been used very successfully in the industry. As a result several major car manufacturers now have firm plans to expand the application of robots to spray painting. As one leading figure in the industry has commented: '(government) requirements and the need to conserve energy will become the impetus to resolving the spray booth's problems simply by removing all the people from the process.'

Fully automatic painting can be expected to become a reality in the near future, and in this particular industry with its regular model changes, this means robots, not painting machines. The advent of other new technologies, particularly in the field of minicomputing (microprocessors), will mean that robots will become increasingly sophisticated in future and lead to the need for less and less human intervention at all stages of the painting process.

The future for robotic painting is so bright that it will develop as a total system rather than as a continued improvement of robots alone. The car industry will demand moving conveyors carrying a mix of body styles and color requirements which will call for proper identification of the unit coming into the painting booth and the provision through computer control of all the demands which that model makes on the total system. Figure 16.2 is a suggested layout for a robotized paint line prepared by an automobile manufacturer, not by a manufacturer of robots.

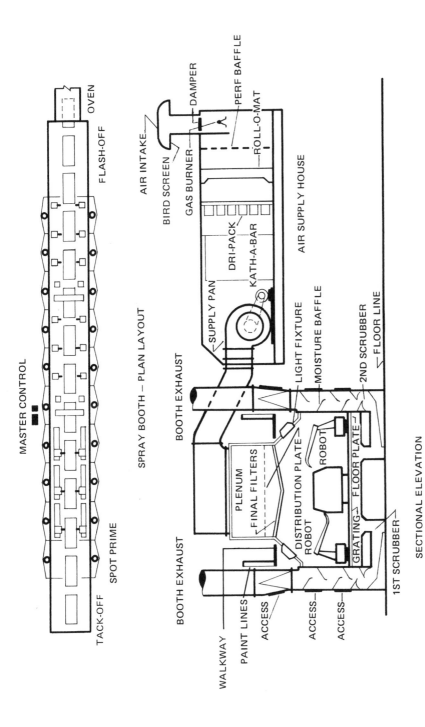

Figure 16.2 *Layout for robotized spray-painting process in automobile industry*

Benefits analysis of robot painting

The following justification for robot painting, also from the car makers, gives some idea of what they expect from their massive investment in this process.

1: ROBOTS WILL ALLOW US TO DEAL WITH A HOSTILE ENVIRONMENT:
- O Noise
- O Carcinogenic materials
- O Particulate matter

2: ROBOTS WILL ALLOW US TO PROCESS WITH LESS ENERGY:
- O Reduced fresh air requirements
- O Reduced exhaust
- O Reduced energy cost

3: ROBOTS WILL ALLOW US TO IMPROVE PAINT QUALITY:
- O Less dirt
- O Uniform build
- O Consistent quality level
- O Cope with specialized spray techniques

4: QUALITY IMPROVEMENTS WILL RESULT IN:
- O Reduced warranty
- O Reduced in-house repairs

5: REDUCED MATERIAL COSTS WILL FOLLOW

6: REDUCED DIRECT LABOR COSTS WILL RESULT

This is quite an impressive list of benefits, especially as it was drawn up by the customer, not someone trying to sell him a robot.

To turn to what is being done now in the automotive industry, Chrysler, Ford and GM are all carrying out experimental painting with the Trallfa robot. Chrysler is developing an 'elephant's trunk' complex with four axes and the ability to paint 60 units an hour, both inside and outside as they move along on a line travelling at 22 feet per minute. The ability of the robot to follow a moving target is one which is of great importance to the car maker.

The robots being used are under microprocessor control and their programs are easily changed. The machine can hold up to 128 minutes of unique paint programs.

General Motors, in its Technical Center Laboratories, is investigating what may become the most sophisticated paint-spraying process in the world.

To meet the demands of what will clearly be a lucrative market, several robot builders other than Trallfa are preparing their robots for this application, and competition may be fierce in the years ahead.

Chapter 17

Plastic molding applications

The use of plastic materials has come about largely since the end of World War II, and in this relatively short period the development of these useful substances has been both diverse and rapid. Today it is difficult to picture a world without plastics, and they are to be found almost everywhere from high-performance aircraft to simple kitchen utensils. Much of the technological development which is taken so much for granted today would have been difficult if not impossible without the availability of plastics to form essential components.

Plastics are mainly hydrocarbons with a peculiar molecular structure. The molecules form long chains or polymers, these comprising chemically linked units known as monomers. The process by which polymers are made is known as polymerization, and the resultant material can show surprising strength coupled with very low density or weight, two useful features for fabricating a multitude of parts. Although there is an almost endless number of plastic compounds which chemists can provide, some of them, such as polyethylene, polystyrene and polyvinyl chloride, are now almost household words — so rapid has been the adoption of these materials into everyday use.

Plastic molding processes

The most useful plastics are a class known as thermoplastics which have the property of softening when heated, only to harden again when they cool. This allows them to be shaped very easily, and it is fortunate that softening and hardening does not significantly alter the properties of a thermoplastic. When heated, a thermoplastic first becomes elastic, like rubber, and then completely fluid like a very viscous liquid. These properties have given rise to a whole new industry, plastic molding. This process is capable of turning out parts at high production rates yet within fine tolerances. There are various forms of plastic molding, each of which has its own particular applications.

Extrusion molding

This process, carried out in a device known as an extruder, is a very common method for shaping plastics. The extruder is a heated pressure-vessel in which is a helical screw like the domestic meat grinder which conveys granules of the plastic material through a die heated to about 200 degrees Centigrade. The shape of the finished product is determined by the die, and a wide variety of shapes can be produced by this technique. For example a circular die will produce rods, an annular die forms pipes, while sheet may be produced from a slit aperture. The emerging section, in continuous form, is finally shaped by passing over rollers cooled by air or water, and the extrusion cut into lengths as required. In such a way would PVC pipes and sheets be produced, the process lending itself to stock materials rather than complex shaped parts which are best made by another method.

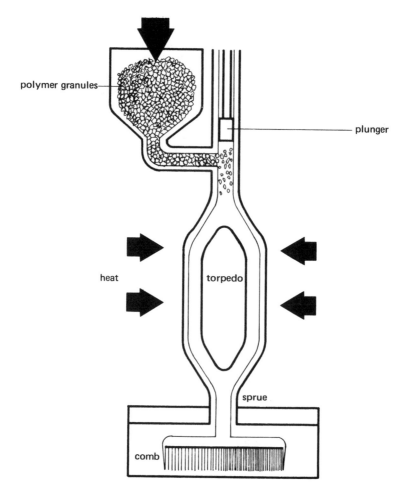

Figure 17.1 *The injection molding process*

Injection molding

This is perhaps the most important plastic molding process. It is capable of producing precise plastic parts in production quantities. Just as in the extrusion process, the raw material is softened by being heated, and conveyed by screw mechanism. The difference is that the screw has an additional role to play, that of a ram. It forces the hot plastic rapidly into a cooled steel mold (see Figure 17.1). Typical operating conditions are 200 to 300 degrees Centigrade and 14,000 pounds per square inch pressure. When cooled, the mold opens (being usually in two halves) and the molded part is removed. This is, of course, a very simplified description of the process; modern injection molding machines are sophisticated and capable of very high repeatability.

Blow molding

This form of plastic molding is restricted to the manufacture of bottles, drums, tanks etc. The raw material is usually a length of plastic tubing, often PVC, which is extruded into a blow mold. As the halves of the mold close over the material, one end is pressed together to close it and the other end connected to a source of compressed air (see Figure 17.2). The tubing is blown up like a balloon until it fits tightly against the sides of the mold which is cooled so that the part solidifies. The mold is then opened and the part removed or, in some machines, ejected automatically.

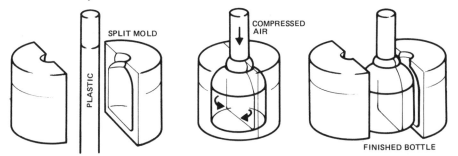

Figure 17.2 *The blow-molding process*

Thermoforming

This method is used to make bowls and similar shaped objects. The raw material is sheet plastic, or film, which is heated, usually by infrared radiation, until it softens. It is then sucked against a suitably shaped, cooled mold using a vacuum technique. The final product must then be separated from the remainder of the sheet material by some suitable method.

Rotation molding

In this rather less common process a hollow mold is used which can be heated, and plastic material, often in the form of a powder, is introduced. The whole mold is then rotated about two axes, somewhat like moving a test-tube continuously in a bunsen burner flame. The plastic material melts and is evenly distributed over the mold. After a suitable rotating period, the mold is cooled, opened and the now-solidified part removed.

Opportunities for robot applications

From the above brief descriptions of the molding process, it will be obvious that the operation has much in common with die casting. In injection molding, for example, the charge of raw material is injected automatically, just as it is in the die casting machine. After the ram operates, it is now necessary to remove the finished part from the dies. If this is done by hand, the operator places himself at risk by putting his hands and arms in the die area. Just as in die casting, as a safety factor, some molding machines are provided with ejector pins which push out the part which then falls into a suitable container.

One of the characteristics of plastic molding is the relatively long cycle time. It is this time which the human operator utilizes to perform secondary operations such as trimming and packaging. These operations are not easily robotized. For example, the trimming of flash is not predictable, and to try to use special dies for this purpose is not cost effective when compared with a human operator using a penknife and eyesight. However robots are already working in plastic molding plants, though they are best suited to dealing with large moldings such as garbage cans. In the larger presses needed for parts of this size the operator is at much greater risk when he physically enters the die space to extract the parts.

Sooner or later the industry will adapt robots on a larger scale because by doing so the process will first of all be better rationalized and because some of the more complex molding procedures such as those requiring steel inserts will lend themselves to higher production rates and better quality when tended by robots. One factor that must be taken into account is the environment. Compared with many of the industrial processes today, the plastic molding shop is relatively benign. The plants tend to be clean and involve no hazards or even unpleasantness for operators.

Current robot use in plastic molding

To date, robots have successfully accomplished the following tasks in plastic molding plants:

○ Unload one or two injection molding machines
○ Trim moldings on removal from machine
○ Load inserts into the mold
○ Palletize the moldings for despatch
○ Package the moldings

It is of interest to look in a little more detail at some of these robot applications.

In one plant a robot is used to unload two injection molding machines making elastomer rubber parts. The specially designed hand enters the open press, strips the part from the die using a combination vacuum and mechanical gripper and removes the part from the press. These are large parts, and when human operators were used they required frequent relief from the noxious fumes which were present in an operating environment approaching a temperature of 400 degrees.

The parts in this application are made two at a time, and therefore it is necessary to separate them. This is accomplished by the robot placing the parts over a cutting blade. The part is then deposited on a conveyor and the sprue disposed of as the robot turns to unload the second injection molding machine. The plant layout for this application is shown in Figure 17.3.

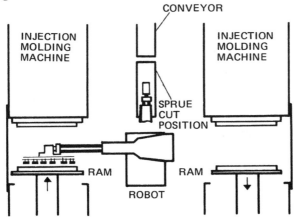

Figure 17.3 *Plant layout for injection molding application*

The special hand for this application consists of a housing containing an air operated mechanical gripper mounted on either side of the housing by two arms pivoted by means of double acting air cylinders. The mechanical gripper grasps the sprue and breaks the center portion of the part from the die. Air jets within the hand directed between the part and the mold help break the adhesion. Figure 17.4 illustrates the hand mechanism.

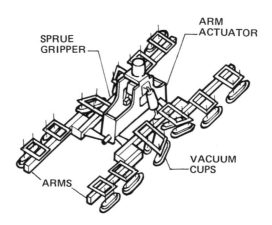

Figure 17.4 *Special hand for injection molding application*

The cost justification for this robot installation comes partly from increased productivity and from faster cycle times and the savings of one operator per shift for each press. Savings in removing operators from the die area of the press amounted to some $15,000, this representing the cost of equipment it would have been necessary to instal to meet the safety regulations imposed by new legislation.

One of the problems ordinarily encountered by robots — that of part orientation — is eliminated entirely in the plastic molding process since the dies determine precisely the position of the part. In Figure 17.5 a robot is seen servicing two machines and utilizing yet another design of gripper, this time equipped with suckers to grip the parts.

In another application a robot has successfully been used to work on a transfer molding installation. In this, a part is unloaded from a machine and then transferred to a second machine where secondary operations are performed, in this case cleaning a large number of holes and grinding part of the molding. The actual movements of the robot are no more complex than operating one machine, but it is interesting to note that by means of random program selection the robot can automatically omit one or more of the transfer molding presses while continuing to operate the remaining presses. This is important when a machine has to be taken out of service without shutting down the entire operation.

It is clear that the robot does not have to carry out any new or unusual operations in a plastic molding application. It is a familiar 'pick and place' job. Consequently the operation of stacking parts in complex patterns presents no difficulty to the robot.

Again, by designing proper hands and grippers it is possible for a robot to load simple inserts into a molding machine before the cycle is initiated. This generally requires two sets of fingers, otherwise the inserts may have to be presented to the robot by means of a suitable

Figure 17.5 *Robot servicing two injection molding machines*

conveyor system.

With the size of injection molding machines on the increase, clamping pressures of 5000 tons being by no means uncommon, the hazards to human operators are considerable and the use of robots is likely to increase dramatically for that reason alone. But there is another factor which has only recently become obvious. The energy crisis has caused the cost of plastic materials to escalate alarmingly so that there are real incentives to lowering the direct labor cost to offset the material increases.

It can take several hours for a plastic molding machine to reach a stable thermal condition and this is essential for the production of uniform parts, both dimensionally and from the point of view of finish. During the stabilization period parts have to be made and they are often wasted. If a human operator is attending the machine during this warm-up phase, extra costs are incurred, but if a robot is installed no direct labor costs are involved although there are of course some costs due to the need to amortize the robot investment during this non-productive period.

Although it is impossible to generalize with any real accuracy, a robot used in a plastic molding application should be able to pay for itself in a year or eighteen months of operation in a typical case.

Chapter 18

Applications in foundry practice

Melting down metals and pouring them into molds to set in the shape of the mold is one of the earliest of man's technological advances. Today this 'founding' process is one of the major manufacturing techniques in industry, and basically it has changed little from those somewhat primitive operations of several hundred years ago.

Strictly speaking foundry practice relates to any melting and pouring operation be it metallic or otherwise, but today the term is generally used to describe those industrial activities where component parts of various shapes, sizes and composition are made by the pouring of molten metal into molds. Mold designs have become very sophisticated to permit the casting of precisely dimensioned parts in clusters to achieve high production rates through multiple casting methods. Foundry work also involves single castings weighing several tons, but in all cases the basic process is similar.

However, in providing human operators with a distasteful environment, foundry operations are among the most offensive activities, with noise, noxious fumes, splashing from molten metal and heavy moving machinery all affecting personal safety, health and the quality of life of all concerned. Yet the foundry industry ranks sixth in the USA on the basis of value added to the raw material by the operations performed on it. Dependent upon it are automobiles, farm machinery, mining, and equipment of all forms.

The casting process

In common with many other industries the need to improve the profitability of foundry operations is now imperative to combat competition, and if this is to be done without further impairing the working conditions for human operators, automation of some kind is essential. This does not represent a new departure on the part of management since the industry has always made use of specialized equipment such as manipulators to lighten the task of the operators.

The work of a foundry is not limited simply to melting down the metal, pouring it into a mold and then extracting the casting from the mold when it has solidified. The complete foundry operation extends to the removal of any unwanted parts of the casting (such as risers, gates and sprues — see Chapter 10 on die casting), and then finishing the part to specified dimensions, usually by a grinding process. Much of what goes on in a foundry requires eyesight, judgement and decision-making based on long experience. These qualities are not readily replaced by automation although much has been achieved with manipulators guided by human operators who use their skills at a more remote point, away from the dangers and unpleasantness while the machine provides the muscle power needed to move a heavy casting.

But even with these aids the foundry is not an attractive place to work, and there is a dearth of good operators in the industry. This has resulted in serious attempts to bring the robot into this workplace with a good deal of success, but there is little doubt that these are early days and the penetration by robots is still to come, as it surely will.

The process of casting involves four main steps; heating the metal until it is molten, pouring it into a mold, removing the charge of metal from the mold when it has cooled sufficiently and finally cleaning up and finishing the part by removing unwanted flash, oxidation products and excess metal. Whether the casting weighs four ounces or four tons, the procedure is the same; only the way in which the operations are carried out differs.

Over the past decade quite significant advances have been made in the technology of casting. Modern electric furnaces are cleaner and permit closer temperature control over the charge of metal. Mechanization of various sorts has been introduced along the molding lines — it should be noted that a single ladle charge may be sufficient to make several castings, requiring the ladle to be moved along a line of molds. Also automation has been applied to the pouring of the metal and subsequent operations.

Molds are generally made of sand. The mold, in two parts which mate together leaving an entry hole for the metallic charge plus some orifices to allow air to escape, is often made under high pressure, so that the sand is very tightly compacted. Very often the surfaces of the mold which are to be in contact with the molten metal are sprayed with a 'die ease' compound, typically a graphite-kerosene mixture. This may be baked on to the surface by raising the surface temperature to about 500 degrees. The use of such a spray is very important in intricate, shaped areas, the purpose being to insure that the cooled casting comes cleanly away from the mold.

When the metal is poured into the mold, good foundry practice requires that the charge flows cleanly into the sprue and that the flow-rate is such that the sprue never overflows. Furthermore, for a good homogeneous casting, the level of molten metal in the sprue should be constant throughout the pouring operation. This is not easy to achieve, especially with very heavy ladles. In some castings, particularly larger sizes, the charge of molten metal is forced in under pressure. Such molds are made up in steel platens, rugged enough to withstand the casting process.

After the charge has cooled down sufficiently, the mold is knocked apart to reveal the casting. With a sprayed mold, one or two sledge-hammer blows may be enough to break away the mold and to clear it of most of the sand and oxidation products which tend to adhere to the metal part. When large pressurized molds are used, however, a shake-out process is required. The charged mold is dropped on to a conveyor system, and the fall starts the break-up of the mold to display the casting within. The casting is carried away while being subjected to sustained vibrations on the so-called 'Vibra Rail', and this tends to loosen any sand and other unwanted material adhering to the casting.

Finally the casting has to be fully cleaned and all unwanted metal cut away, usually by a grinding tool or with a flame cutter. The excess metal will be not only that which is squeezed out around the periphery of the mold but also the sprue, gates and other parts of the mold which are included in it to facilitate the casting operation.

This cleaning up is known as fettling the casting. Since castings may be of quite complicated shapes, fettling is a time consuming operation.

Robots in the foundry

Because it is such a noxious environment, the foundry seems an obvious place for robots to find useful applications. So far the penetration of robots has been slow. First of all, although the industry is geared to automated devices of various sorts, these tend to be less sophisticated than in some of the industries where robots have made their mark. The ladle of molten metal may weigh several tons, or it may be small enough for one man to pick up easily. Pouring is something of an art, though the repeatability of robots may change this in the future. The charged mold will seldom be orientated in such a way that a robot can grasp it reliably, and the space required to move a large casting around during the shake-out and trim may be too much for the robot's reach.

These are all negative comments, so what in fact has been accomplished by robots in this application? First, ladling; robots have been used for some time in cold die casting machine applications to ladle the hot metal into the piston section of the machine prior to charging the die. Ladling is also possible in foundry work depending upon the size and weight of the ladle and its charge of molten metal. At the present time the applications are few and restricted by size.

One effective application is to be found at a company on the west coast of America. The company, California Wheel, uses a robot to pour aluminum into molds forming wheels for the automobile accessory trade. Since the furnace is relatively small, and the ladle both positively orientated and manageable, the robot picks up the charge of metal on receiving an appropriate signal and swings round to the mold, also registered properly so that the mouth of the ladle comes over the sprue as the robot tilts it downwards. By means of its program, and using time as the control parameter, the robot returns the ladle to a horizontal position when the correct charge has been poured into the mold. Considerable consistency of product results.

Next, mold care; just as robots are used to spray the die faces in a die casting machine, so they can accurately impregnate the surfaces of a sand mold provided that the mold can be guaranteed to lie in the correct position with respect to the robot. This application is used in several foundries today.

Baking the 'die ease' coating on to the mold after it has been sprayed is an obvious follow-up operation for the robot. This is done by having the robot hold a torch in its hand which it passes over the surface of the mold following a car fully programmed pattern. Surface temperatures of around 500 degrees are demanded, and again the robot times the operation, changing what was once an art performed by a human operator into a routine machine operation.

Robots are at home in unloading various types of machine. After a casting has come through the shake-out process, the robot can grasp it and carry out further operations on it if the part can be orientated correctly. One of these further operations is to cut away the excess metal. By providing the robot with a cutting torch in its hand, it can cut through a 50-pound casting in about 10 minutes. It is significant that this flame cutting sequence is one of the dirtiest in the foundry, so the abilities of the robot in this application are welcomed by foundry operatives. The robot can go a step further, however, for having got rid of such tangible parts as the sprue, it can grasp the part and place it in a trim press which will remove other unwanted metal.

Figure 18.1 shows a robot used for cutting risers from steel castings and for general fettling duties in the Kohlswa Steelworks in Sweden.

Applying robots to the fettling operation

The degree to which a robot can be used depends on a number of factors, and as has been stated, it will not be suitable for very large castings nor for any operations where there is too much randomness in the positioning of parts being presented to the device. Where robots are being used they are part of a man-manipulator-robot team where each plays a part. The aim must be for better working conditions and higher production rates. When there is a significant saving to be made in both areas, it may pay to go to lengths to accommodate the robot by modifying the layout or the procedures used. One area where this looks very promising is in the fettling shop. The fettling of castings has

Figure 18.1 *Robot trimming steel castings at Kohlswa Steelworks, Sweden.*
By courtesy of ASEA

changed very little in the past 30 years. There is a very good reason for this. It is not uncommon for a foundry to cast small parts in batches of 10,000 or more. To clean so large a number mechanically is a very difficult task, and so traditionally human operators have done this job using their dexterity and ability to see the product in judging how to achieve the desired result.

As much as 20% of the running costs of a foundry can reside in the fettling shop. The job is by no means a pleasant one because of the associated noise and dirt which demand the use of protective clothing, making the operator feel very hot and stifled. This has resulted in absenteeism being higher in this shop than in other parts of the foundry. It is also difficult to recruit sufficient labor to man a fettling operation, so, all things considered, this appears to be a very fertile area for the robot to take root. The robot is suggested advisedly since an average foundry makes so many different castings that any form of special purpose automation would be hard pressed to cope with the variety of shapes and sizes which have to be fettled.

While this concept is very akin to other uses to which the robot has been put, there is a difference. In fettling, the handling device, whether it be man or machine, encounters severe dynamic loading as the casting is held against the cutting tool. Since significant amounts of metal are removed the result is more like a machine tool than a handler, although the final machining accuracy is of a lower order.

Although industrial robots are not yet being used to any great extent in the fettling process, extensive work in Sweden, Japan and the United Kingdom has indicated that this is a real opportunity for the robot. Its advantages are its large muscle power and the ability to work for hours on end in the bad environment of the foundry. The limitations which they suffer are mainly their lack of dexterity and slow speed when compared with a human operator. This book stresses time and time again that robots have no sense, and this is a serious disadvantage, even in as simple a task as removing flash, since the amount to be trimmed off can vary considerably from casting to casting. To overcome this problem, the cutting tool can be mounted on a constant force suspension system which will take up any variability and keep the cutting edge in contact with the workpiece right up to the stop position determined by the final contours of the part being fettled. Another, but more complicated way of approaching the subject is to equip the robot with some sensory feedback so that the robot's control system can make adjustments accordingly as the amount of metal removed brings the casting closer and closer to its final shape.

In Figure 18.1 the robot is seen equipped with a bracket on which is mounted a powerful electric motor and a large diameter cutting-off wheel.

If the available muscle power of a robot is to be used effectively, a high-power cutting tool is essential. High power in this context is something in the region of 10 kilowatts, a much higher power than is encountered in normal portable tools. This suggests that a better arrangement is to fix the machine tool and use the robot to hold and move the casting against the cutting edge. This permits the use of fixed 20-25 KW grinders which are already standard in many fettling shops.

Having decided on fixed versus movable cutting tools, the next consideration is how to present all faces of the casting to be cleaned to the grinder. Whether the robot holds the casting or the cutting tool its dexterity is likely to be severely challenged with many of the casting shapes which are commonly made today. Many require a full 360 degree rotation for complete cleaning which is a tough specification even for a full six-axis robot. Work has been done on a rotational table so that by moving the part in relation to the robot the need for such an extensive rotation is removed.

Much of this work can be attributed to Dr. Brian W. Rooks formerly of the University of Birmingham, England. His conclusions, after a long experimental investigation of fettling problems are as follows.

1 Much of the output of castings produced can be fettled using robot technology which is currently available.

2 The main requirements are the attachment of power tools of high power rating to the robot arm, or to grip and manipulate castings through a full 360 degree rotation.

3 If the casting rotation method is used it has the advantage that much more powerful tools can be used which increases the rate at which metal can be removed and therefore achieves higher production rates.

4 The capital investment in robotized fettling is very high since it includes auxiliary equipment to convey the parts to the robot, orientate them correctly and present them to the robot. With the large variety of castings typically encountered, this auxiliary equipment needs to be as adaptable as the robot, a not easy matter to accomplish. This problem is less when very large batch production is envisioned, but even special purpose machines may be limited as to the size and shape of the casting.

5 There remains an urgent need to provide fully automated fettling in foundries because of the bad environmental factors. There is no doubt that the robot will play an increasingly important part in such mechanization and the abilities and limitations of the robot will to a large extent determine the design features of machines developed for fettling purposes.

Chapter 19

Machine tool loading applications

None of the industrial processes described so far in this section are capable of shaping materials to the fine tolerances and smooth surface finish which are essential for the component parts of mechanisms like the airplane, the automobile engine, the domestic dishwasher and the host of other devices which are an accepted part of the technological society of today. Such components are fabricated by machine tools.

In this family are the lathe, the milling machine, the drilling machine and the precision grinder, plus several other variants of these basic tools. They are characterized by the fact that they mostly use spindle-mounted cutting-edges, the spindles being in such solid-based and rigid bearings that the tools are 'stiff' and do not give unduly under the pressures exerted by the cutting edge on the workpiece. Consequently high dimensional accuracy of the machined part is possible, and equally good repeatability. Final dimensions to within a few thousandths of an inch are commonplace today. Yet the machines are not new. Watch and clock makers used empirical lathes hundreds of years ago and with them achieved remarkable precision considering the rudimentary nature of their equipment.

Development of automation in the machine shop

Groups of machine tools make up the well-known machine-shop or jobbing shop, the backbone of the manufacturing industry for as far back as records exist. Until recently machine shops were the domain of skilled artisans who used their experience coupled with hand measuring tools such as the micrometer and caliper to work to incredible accuracies.

Today the picture has changed somewhat although it will be a long time before the skilled operator is no longer needed. But the signs and portents are emerging. The following news clipping was taken from the New York Times News Service under a Houston, Texas dateline for January 1979.

> In Japan, a new era of unmanned, fully automated factories, so flexible that they can readily be reprogrammed to turn out different products, is in the offing, and far in advance of similar developments in other countries.

The general opinion in the USA gives credence to the Japanese lead in the objective of achieving unmanned manufacture. But, the Japanese are not alone on this frontier and it's everyone's game. Much rides on such advanced machine tool control technologies as N.C. (numerical control), C.N.C. (computer numerical control) and D.N.C. (direct numerical control). We will return to the potential of N.C., C.N.C., D.N.C. and other exotica in due course, but right now the thrust is toward labor saving by robots.

Ever since products of any sort have been made the demand for greater production rates (and by implication higher profitability) has been a dominant issue. Since labor content has always been significant, attempts have been made from quite early times to reduce this element by the introduction of techniques which do away with the need for human operators. Thus automation of the machine shop has been steadily on the increase for years.

The objectives of automation in the machine shop, as elsewhere have always been to reduce in-process inventory, cut lead time, minimize direct and indirect labor costs, make maximum use of capital equipment and generally increase the number of acceptable parts made per shift. Often these aims had to be pursued in the face of regular product changes which made general purpose automation difficult to apply. One system which grew out of these developments was the transfer line.

In a transfer line the intention is never to let the part leave the line but to keep it moving in indexed steps from operation to operation. A limitation of the system is that the line can therefore move only at the rate determined by the longest machine operation. Another disadvantage, already mentioned in the earlier chapter, is the very specialized nature of the line which can obsolete it before it has generated a reasonable return on the necessary investment.

But what is the role of robots in this realm of the manufacturing industry? What more obvious role could there be than loading and unloading machine tools by a robot, designed to displace human beings in the more rudimentary tasks in industry?

At first sight it seems to be a simple enough task to pick up a part, put it in the chuck, wait until the machine has done its work and then reclaim the machined part, and if necessary pass it on to a secondary operation. Even better, if one used general purpose machine tools in conjunction with industrial robots, machining centers could be created which were much more flexible than conventional transfer machines. Also, since these centers would be using off-the-shelf machines and non-specialized robots, they would cost a great deal less to instal and run, come on line very much faster and be free from the bugbear of built-in obsolescence. However this beautiful picture was met with profound apathy on the part of potential users, and thus it was not in this field that robots found their first jobs but instead they gave yeoman service in die-casting, welding, forging, plastic molding and many other applications. Only recently has the idea of robots tending machine tools been received with enthusiasm, and this new acceptance was born of tough economic pressures. Even the most ultraconservative manufacturer is now forced to admit that robots have come to stay, after 15 years or more of growing penetration into the industrial workplace.

The introduction of N.C. machines has helped a great deal. Those who have adopted N.C. in their shops find that robots are not so dissimilar, and they feel comfortable with this type of technology. After all, both N.C. machines and robots are controlled by taped programs and operated by electrical signals generated both from within the program itself or from an external source synchronized with the movements of either machine or robot.

However, mere technological achievement never swayed the successful manufacturer, but profits would! Since the first robots entered industrial service in 1961, labor costs have increased by no less than 250%, whereas the cost of owning a robot has gone up only 40% — and of course the robot of today is vastly more sophisticated than the 'Model T' of the early sixties, functional though it undoubtedly was.

The attitude of the machine tool builders has to be taken into account too. They naturally preferred to keep the design work for special loaders and unloaders in house. It was good business once, but if anything, design costs have outstripped direct labor costs, and customers are realizing that a series produced robot is a better bet, economically, than a purpose built part handler. The fact remains that robot technology today is powerful enough to overcome most of the early objections, whether they were simply specious or truly substantial.

Robot applications to machine tools

There is already a classical layout for using robots in machine tool applications. Figure 19.1 is a fairly typical example. The robot stands stage center surrounded by the machines it tends.

Metal cutting, especially when the parts are large and the volume is low to medium, is a lengthy process which means that the part spends a long time in station. If the cutting cycle exceeds about 20 seconds and particularly if the time of dwell runs into minutes, it does not make economic sense to have the robot unoccupied while the machining cycle is completed. Secondary and tertiary assignments for the robot should, in such cases, always be sought.

Figure 19.2 illustrates a robot stationed among three machine tools. This particular installation has been producing completely machined valve bodies on a two-shift basis for over four years with high reliability.

Without stretching the analogies between human operators and

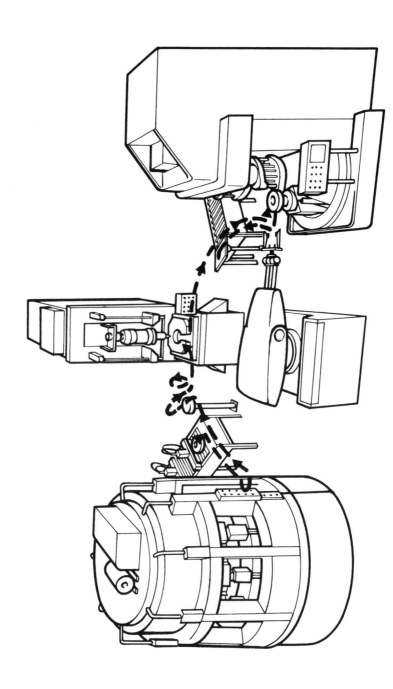

Figure 19.1 Typical layout for applying robots to machining applications

Figure 19.2 *Robot tending three machine tools*

robots too far, it is a fact that when a single operator tends several machines, each with long cycle times, he must walk between them and keep them all going. When machining cycles are particularly long, the robot handling workpieces can be made to travel among more machine tools than can conveniently be placed around a stationary robot. This system is discussed in the concluding 'Robot travels to work' section of Part I, Chapter 4, and illustrated in Figure 4.7. The robot depicted handles no less than eleven different machine tools and even carries along with it a buffer station for parts in intermediate stages of completion.

It can be noted in passing that the robot of today likes some room to swing its arm, and therefore to adapt an existing transfer line to robots may be difficult. Transfer line machines are usually put as close together as possible; the moving robot may therefore be the only solution if the factory layout does not permit of sufficient room for wider spacings to accommodate stationary robots. However the final design is usually on economic grounds, and if one moving robot can do the job of three or four stationary ones the technique speaks for itself.

For a further illustration of the travelling robot in the machine shop, the reader is again referred back to Chapter 4, Figure 4.8, which shows an overhead mounted robot moving along rails to service eight N.C. lathes. In fact the installation shown is a very up-to-date one, with full computer control. Both the lathes and the robot take their instruction from a central computer. In the control room is a library of tapes to program the lathes for a number of machining operations as well as providing instructions to the robot to load, unload, move etc. The robot movements are 'choreographed' by the central computer along the line (which is fully 200 feet long) to minimize lathe downtime. The similarity to a supervisor walking up and down the line keeping things going is undeniable! The system is illustrated in schematic form in Figure 4.9 in Chapter 4.

All of the installations described thus far are to be found in Japan. According to recent estimates, the Japanese have more robots in action than any other nation. Of these, they most definitely lead the way in machine tool applications which lends support to the suggestion voiced in the press cutting – that soon Japan will be operating completely robotized automatic factories with the ability to change products by changing computer and robot programs.

In the USA, Xerox Corporation has a high volume line using three robots, a transfer conveyor, two center-driven C.N.C. lathes with double-end capability, and a supporting cast of brazing, grinding, broaching and turning machines. The layout is as shown in Figure 19.3. The three robots have a 10 feet reach and are used to transfer the parts

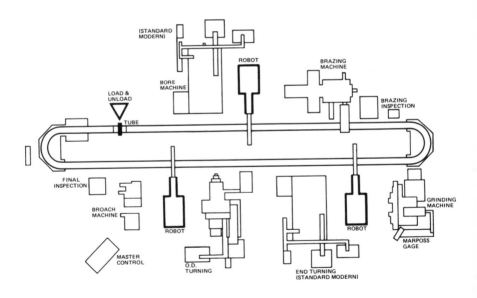

Figure 19.3 *Layout of three-robot line in machine shop at Xerox Corporation*

between the conveyor and the machine tools. Nine basic programs will accommodate the entire family of parts made, while alternate bypass programs come into operation during any machine tool downtime. These programs are on cassettes to permit rapid product changes to be made. To oversee the whole operation, a supervisory programmable controller takes care of automatic program selection, the synchronizing of robot movements with other line operations, cycle initiation, fault monitoring and emergency shutdown. Each of the three robots has two hands which can be used independently to clamp and unclamp the parts. This speeds up the handling of finished and unfinished parts at each operation, keeping the load and unload cycle time to an absolute minimum. In Figure 19.4 one of the two-handed robots can be seen loading a lathe.

Massey Ferguson produces four different sizes of planetary gears in a schedule in which the volume can vary from low to high. The company was contemplating a transfer line but decided instead to opt for the greater flexibility of a robotized line. The installation which resulted is shown in schematic form in Figure 19.5. Each of the three robots has its own work station and each performs a different operation. Conveyor transfer and inter-station buffer storage are conventional.

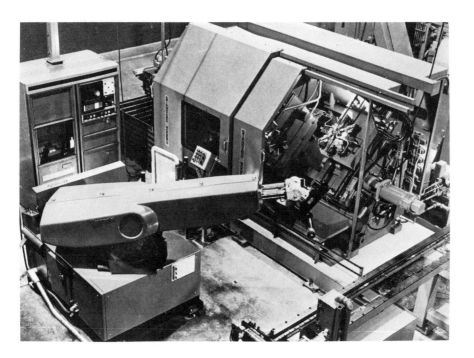

Figure 19.4 *Double-handed robot loading a lathe*

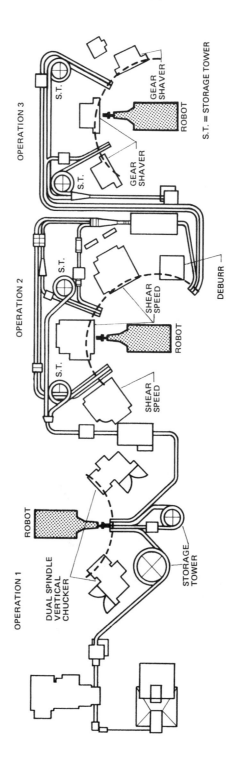

Figure 19.5 *Layout of three robots on machine line at Massey Ferguson*

Some idea of the flexibility achieved can be gained from the fact that each of the three robots can search for availability among its own machine tool complement to keep the line working during random downtime or scheduled tool changes. Willing workers indeed!

Figure 19.6 shows the supervisory controller which monitors the entire operation. For ease of interpretation, the panel of this monitor is laid out as a schematic diagram of the actual robotized line. Lights flash where the action is (or isn't), leading the eyes of maintenance staff right to the spot.

Figure 19.6 *Programmable controller used with triple robot installation at Massey Ferguson*

A robot transfer line has a considerable edge over a hard automation line and compared with manual operation, production is 25% faster. The complete system payback in this case is estimated at 2¼ years. This is good news for the large scale, big company manufacturer. But what about the small operator who has to contend with batch manufacturing? Well, he is not out of the running, even at this early stage of robotized machine tools. A recent development, aimed just at this problem, is to integrate a robot into a system of conventional N.C. equipment as shown in Figure 19.7. With the concept comes a parts classification system which will be very useful in work-loading the system. Again the double hand is much in evidence (Figure 19.8).

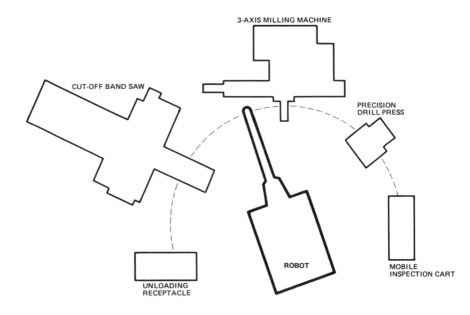

Figure 19.7 *Integrated robot-N.C. system for small batch manufacture*

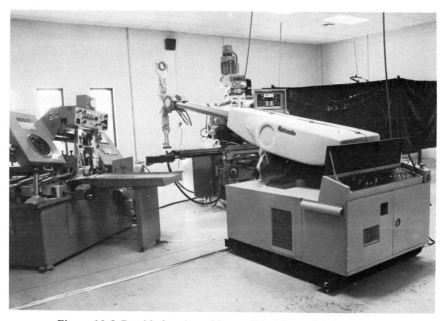

Figure 19.8 *Double hand used in small batch machining system*

Robot attributes for machine tool applications

Part I of this book has already dealt with some existing robot attributes and those which are desirable for the future. In machine tool applications (just as in any other role) it goes without saying that the robot must have a long enough reach to service all work stations and be able to carry the heaviest parts being handled by the system. The following check list spells out other desirable features, some being essential while others would simply make the job a great deal easier.

1: UP TO SIX INFINITELY CONTROLLABLE ARTICULATIONS BETWEEN ROBOT BASE AND GRIPPER APPENDAGE

Sometimes it seems that three articulations ought to be enough to load a machine tool chuck, but the real world finds otherwise. Often manipulation patterns are complex to avoid machine structure. The use of dual grippers compounds motion requirements and the requirement to lend a variety of machine tools may demand great variation in load-unload paths. And palletizing speaks of a variety of arm destinations.

As to making the articulations infinitely controllable, one can only note that the 'limited sequence' robot rapidly runs out of useful arm destinations.

2: FAST 'HANDS-ON' INSTINCTIVE

In a machining system, it's nice to do a layout knowing that anyplace reachable by the robot can be quickly programmed on the job. If a range of parts must be handled, all programs should be easily generated and stored for use as needed. This is most easily done using 'record playback' instinctive teaching with actual parts at hand.

3: REPEATABILITY TO 0.3 mm.

Quite a few raw castings and rough machined parts can be given final orientation by ingenious centering grippers, but still, accuracy is necessary to center the workpiece into the machine tool checks.

4: SPEED EQUIVALENT TO THAT OF HUMAN OPERATOR

If a robot is slow on the job, the machine tools will not be utilized at optimum capacity. A loss in productivity can eat up economic benefit derived from human operator replacement.

5: PROGRAM SELECTION CAPABILITY FOR CUT-OUT OR ALTERNATE ACTION

Alternate action is very commonly needed in a robotized machining center. A machine being serviced will demand that arriving parts be buffered. Perhaps any of a group of identical machines must be bypassed on occasion. If an inspection station is included, workpiece destination will depend upon commands from this station.

6: COMPATIBILITY WITH WIDER NC SYSTEMS

There are sophisticated systems that control machine tools in the DNC or CNC mode and they are under continual command changes. The system's robot should be able to respond with equal alacrity.

Sometimes, it is possible, at cost advantage, for a robot to share an NC controller that has spare capacity and a compatible interface.

7: PALLETIZING AND DEPALLETIZING CAPABILITY

In many applications, it is not practical to present workpieces one at a time in a single location. To avoid human attendance, parts may be delivered on a pallet to the work station, and the robot may also palletize the output. Palletizing and depalletizing capability provides for inter machine buffer storage so that a system can stay on stream when machine elements are temporarily down.

8: LOCAL AND LIBRARY PROGRAM ACCEPTANCE

As long as a robotized machining center is designed to handle a range of parts, then a range of programs will be necessary for the robot or robots. Some of these programs must be on hand for instantaneous recall, others may be extracted for external storage and later convenient introduction into the local memory.

9: HIGH RELIABILITY – NOT WORSE THAN 400 HOURS MTBF

A robot tending a machine tool – or worse still, a group of machine tools, threatens the system with a 'domino effect'. If the robot goes down, the system goes down and production is lost. Human operators could stand in, but that defeats the labor saving purpose.

The answer is high reliability, better than 400 hour MTBF, mean-time-between-failure. If the robot also has a low MTTR, mean-time-to-repair, then it can be more dependable than human labor. Two percent downtime is a reasonable demand, outstripping human downtime which now runs about 3.5% in U.S.A. metal working.

10: DUAL GRIPPER CAPABILITY

The dual gripper capability is related to point 4, speed. With dual grippers lost motion is avoided, since the robot arm can strip a completed part and load a new part without leaving the machine tool bed. Naturally, point 1 figures in, because manipulative power is essential in handling two hands mounted at the end of one arm.

11: MOBILITY IN WORK AREA

As we have seen, there are applications where a single robot can tend to more machines than can be conveniently arranged around a stationary robot. Then, the robot should be mobile. Human flexibility is not necessary; only the ability to move linearly, as on tracks, among work stations.

12: AUTOMATIC SENSING FOR CHUCK ALIGNMENT

Often, parts do not have symmetry around the chuck center line and the parts must be rotated to random locations to permit engagement. A special robot gripper sensing mechanism performs this task. Better still, newer machine tools provide for fixed chuck destinations so that engagement need not be sought by the robot.

13: ADAPTIVITY – INCLUDING RUDIMENTARY VISION

Adaptivity can cover all of the capabilities that a human brings to the task. Use of instrumentation can help the robot to react similarly. So it is with tool wear and breakage and finished part inspection. Scrap accumulation, where chip breaking is inadequate, may require rudimentary vision. That's on the way, and a robot worth its salt will accept vision module inputs.

Robotics and NC are obviously complementary in control technology, but they are also part of a larger manufacturing technology scene that is inexorably driving industry toward the goal of unmanned manufacturing. Impetus also comes from workers who rightly rebel against debilitating factory jobs and exact ever-increasing 'hazard pay' to push up manufacturing costs still further.

Chapter 20

Heat treatment applications

Heat treatment is a generic term covering a variety of processes carried out on metals and metallic alloys to modify their properties in some way. Such procedures, as the name suggests, are aimed at heating the material to high temperature and then cooling it. Sometimes the cooling occurs normally by removing the heat source and allowing the part slowly to return to ambient temperature; at other times the treatment may call for the part to be quenched in some suitable liquid while still hot. The aim is always to produce properties suitable for a particular requirement. The principle is that the atomic structure of the material can be modified by a controlled heating and cooling process. Typical characteristics which are affected by heat treatment are hardness, strength, ductility and electrical characteristics. Heat treatment may be necessary to prepare a metal for machining.

Heat treatment processes

The basic ingredients of a heat treatment operation are a furnace in which to heat the part, and a medium in which to cool it again. The essential parameter is time, and the duration of both the heat cycle and the cooling cycle are crucial to the result obtained. Much of the technique depends on experience since the variables involved are not easily calculated from first principles.

One of the metals most frequently subjected to heat treatment is steel. This metal has a very wide range of possible chemical compositions which, in turn, govern the properties of the metal. Three main heat procedures are carried out on steel; they can also be used with other metals though as already stated there is a great deal of know-how involved.

Annealing

This process is for softening steel so that it can be worked more easily. However the technique can be used to remove stresses from the metal and produce a more homogeneous structure. For this reason parts such as forgings, castings and cold-worked samples are often subjected to annealing.

Annealing is carried out by heating the sample in a furnace to a critical temperature above which the substance *austenite* is formed. Austenite is a non-magnetic solution of ferric carbide (or carbon in iron). If the sample is then allowed to cool slowly in the furnace, the structure is converted to a form which can easily be machined. The reasons are complex, and depend on achieving the correct mix of ferrites and carbides within the molecular lattice. The process is referred to as 'Aus-forming' in industry.

In another form of annealing, the material is heated for a long period (or 'soaked') at temperatures just below the critical level for the material being treated. In this case austenite does not form, but instead a ferrite-carbide structure somewhat spherical in structure evolves, and this proves to be easier to machine than when the full annealing (heating above critical level with slow cooling) is performed.

The process clearly has many variations but all consist of heating the material to a predetermined temperature and then allowing it to cool, usually at quite a slow rate.

Tempering and hardening

If steel is heated to the range of temperatures at which austenite forms, and then cooled rapidly by quenching in a tank containing some suitable liquid, a substance known as *martensite* is formed which has hard and brittle properties. Typical quenching fluids are oil, water and brine. Other factors influence the hardness of steel of course, notably the presence in the lattice of such metals as chromium, nickel, molybdenum, tungsten and vanadium. Such steel alloys will harden simply by being treated and then allowed to cool in air, this again emphasizing that the whole subject of heat treatment is based on long practical experience.

After a sample of steel has been hardened it is common practice to remove some of its brittleness by a process known as tempering. Tempering gives the metal a degree of toughness, though this is only obtained at the expense of some of the hardness of the sample.

In hardening, the material is heated to temperatures between 700 and 900 degrees Centigrade according to the carbon content of the sample. If, after hardening, the material is then soaked at a somewhat lower temperature, say 400 to 700 degrees Centigrade, the structure of the material changes in such a way as to give toughness (that is, less brittleness) with a good degree of hardness. Such materials are almost impossible to machine, though they can be surface-ground. The road-springs of an automobile are typical examples of the need for hard, tempered steel in an everyday application. Gun barrels and wrenches are other well-known uses for such materials.

Surface hardening

This explicit title refers to a process which produces steels with a very hard surface and good overall strength. The need for such a material arises whenever the surface of a component is subjected to sustained wear. Surface hardening can be accomplished in two main ways. The first is to heat just the surface layers of the material, either by holding the sample in a flame or by using an induction heating furnace. When the surface temperature reaches the range at which austenite forms, the material is quenched and a hard shell forms around the outside of the sample.

Another technique, applicable to low-alloy steels, is to change the surface chemistry by the introduction of carbon and nitrogen. This is done by heating the sample and then dipping it into a flux of the appropriate chemical compound.

Since heat treatments promote oxidation of the sample, it is often necessary to carry out the process in a controlled environment from which air is excluded. Such environments can be gaseous or solid, the latter often being in the form of a pack of molten salt surrounding the material being treated, although a solid environment of other chemicals may be used when surface-hardening is performed. It is this pack which provides the necessary compounds needed for the surface-chemistry effect.

Non-ferrous alloys can be annealed also, especially brass and copper, but it is rare for such alloys to be capable of being heat-treated to improve their hardness. There are, however, other ways of hardening some non-ferrous alloys by a process called precipitation hardening, but that is outside the scope of this section.

Heat treatment — aus-forming, hardening, tempering — depends on temperature, time and medium. What, then, are the jobs for robots in this industry within an industry which is so important in present-day manufacturing technology?

Robots in heat treatment

From the range of robot applications described thus far, it would appear that there is a real chance for these hard-working machines to enter this field, and indeed they have done so.

In die casting, forging, press work and molding, the robot has proved that it can thrust its hand into a die, a furnace, or the like, with high precision. It can place a part accurately or grasp one and move it somewhere else. These are all steps in the heat treatment process, added to which the robot will not feel the heat of the sample which it picks up and transfers to a cooling bay or to a quench tank.

No special features are needed for a robot to work in this application;

it is a 'pick and place' job, governed by synchronizing signals linked to the timing apparatus which determines the duration of the heating and cooling periods. Orientation of the part, at least for the unloading of a heat treatment furnace, is usually no problem, though steps may have to be taken to present the part to the robot if it is to load the furnace as well. There is a bonus to be earned too. Heat cycles are relatively lengthy affairs, so the robot has time on its hand. Rather than waste this, it can be put to other tasks during its free time, provided that the factory layout is such that the robot has the necessary equipment within reach.

In the Caterpillar Tractor factory, a robot picks up a pin from a magazine feeder and inserts it into a groove in the rotary hearth of an induction heating furnace. The use of a groove provides an accurate registration for later pick-ups. After placing the pin, the robot picks up a hot pin, already in the furnace, removes it, signals the furnace door to close, and then places the hot pin in a quench tank. The whole cycle described takes less than one minute. By employing electrical interlocks, the robot is prevented from loading a pin into the furnace until it has been signalled that the hearth has rotated to accept it and the furnace door is open. Ejection of pins from the quencher is automatic, but the robot cannot load a pin into the quencher until it receives a signal telling it that the previous pin has been removed.

In another installation at a different factory, a robot is busily operating a heat treatment and quench cycle on parts weighing 40 to 50 pounds. These are fan-shaped blades which are finally assembled into a helix for the inside of truck-mounted concrete mixer drums.

Before the days of the robot on this job, the blades were cut manually from steel plate by a rotary shear, and then cold-formed. Next they were transferred to a furnace area where gas torches arranged in a specific pattern heated the areas to be hardened. After heat treatment, the parts were quenched.

The furnace in this installation is arranged in four levels. After it has picked up a blade and placed it into a programmed position in the furnace, the robot unloads the blade that has been longest in the furnace, then triggers the door which closes. The heated blade is then placed, by the robot, into a 20-ton press. As the robot withdraws from the press die-area, it signals the press to close on the blade to shape it. As the press concludes its operation and opens, this signals the robot to remove the blade, place it at a discharge point and then to go back for another blade to be loaded into the furnace at the commencement of another complete cycle.

This cycle gave the robot some spare time, so the company fixed up a punch press making a completely unrelated part and had the robot load it from a nearby magazine, and, after the press had operated, unload it into a tote box. The introduction of the robot in this factory increased production, simplified quality control, and reduced both indirect labor and overhead costs.

At a Canadian plant operated by International Harvester, harrow discs were heat treated to toughen the part so that it could better resist breakage when it struck a rock in use.

This process was very unpopular among workers in the factory; it was hot and the parts were heavy. A decision was made to save three men *per shift* by introducing three robots as star performers in a well-planned layout. As the system now operates, robot no. 1 stands at the entry conveyor system for the furnace. Its hand is equipped with a vacuum cup which comes down vertically from the robot arm. The hand lifts the top disc from a palletized stack of about 50 discs, and then transfers the disc to the conveyor which carries it into the furnace operating at 1650 degrees Fahrenheit. As an interesting and effective detail, if the robot hand finds no disc in the stack, the vacuum cups rest on the framework of the pallet and a pneumatic pressure sensor terminates the program while a new stack of discs is advanced into position. When this is done a signal restarts the program.

Robots nos. 2 and 3 along the line are each equipped with a two-fingered hand which grasps the disc on its outside diameter. Since the factory makes discs of five different sizes, it is interesting to note that only the fingers have to be changed when a different diameter disc is scheduled.

The ability to switch programs rapidly on the robots is another advantage when the size of discs is changed. A run of any one disc size lasts an average of eight shifts, but on the other hand it may go on for a week. However the robots can change programs in a matter of minutes. Only three-axis robots are needed to perform this relatively simple sequence of movements. Figure 20.1 shows the production line. As each disc emerges from the furnace it continues on the conveyor to a 'pop-up' station which positions the disc for pick-up by robot no. 2. A gating system holds subsequent discs until the robot has picked up the first one and the pop-up station has dropped back. The robot then moves through about 160 degrees and loads the disc into the die of a press. Before the robot hand enters the die, three key conditions must be met:

1 The die is open, indicated by a ram limit-switch.
2 Robot no. 3 has signalled the removal of a previous disc (now dished through the press action).
3 An infrared scanner indicates that the disc is within the required temperature range (450 to 600 degrees Fahrenheit) to be properly formed by the press.

If the disc shows itself to be too cool, robot no. 2 switches to a reject program. A too-high reading indicates that two discs have stuck together

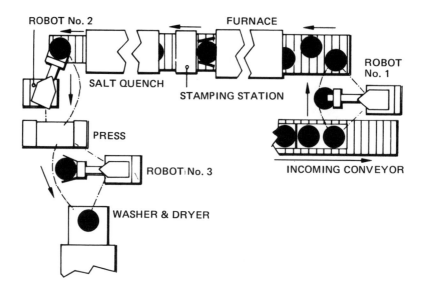

Figure 20.1 *Plant layout for robotized heat treatment line*

and that they are rejected also. Robot no. 3 unloads the press and swings some 180 degrees to place the disc on to the entry conveyor of a washing and drying unit. This is necessary because during its journey through the furnace the discs encounter solid quench in the form of salt, so they emerge heavily contaminated with this compound and it must be cleaned off before the discs can go into service. In this particular application, however, the salt build-up on both the press dies and the robot's fingers is minimized by the use of compressed air jets raised to furnace temperature and which play on the vital parts.

An analysis of the effectiveness of this robot installation is worthy of consideration. The plant's maintenance superintendent made no claims for a higher production rate when the robots moved in. About 300 discs an hour was the rate achieved and this was no more than could be expected from human operators. What was significant was the fact that previously there had been days when production could be maintained for no more than five hours. With the increased reliability of the new line and the indefatigability of the robots, production now can go around the clock.

The decision to instal robots resulted in improved product quality, increased volume, savings in product cost and less labor costs from a reduction in workforce of between six and nine people according to the number of shifts worked. His assessment was not all starry eyed however. All-in maintenance costs with the robots in the system were higher than during manual operation, but considering aggregate savings, this was not incapacitating.

Chapter 21

Applications for deburring metal parts

Metal parts made by mass-production machining techniques almost always contain burrs, that is, irregularities in the form of sharp metal fragments attached to the machined surfaces of the part. Removing such burrs is an expensive yet necessary operation. The most common way by which this is done is for manual workers equipped with suitable small tools to inspect and clean up every part, an expensive and time-consuming process, especially when the burr is in an inaccessible position such as at the end of a blind hole machined in the part. There are some simple deburring operations which can be carried out by sand-blasting or tumbling methods, but the majority of situations are not amenable to any treatment other than careful hand-tooling.

The location and form of burrs are seldom predictable with any real accuracy. While it can be assumed that a particular part will always come off the machine with burrs, the shape and size of the burr can differ widely between parts which are otherwise essentially the same. Additionally, during machine-tool fabrication, the tools suffer wear and tear so that the burr formation gradually changes, then alters abruptly again when the cutting tool is changed for a sharper one.

Demands of the deburring operation

If a repetitive mechanical method of removing burrs is to be used, then every part must be subjected to a deburring operation whether a burr exists at a particular point or not. Otherwise careful visual inspection would be needed to select those parts requiring attention, and this would in most cases be uneconomic. Similarly in a mechanized deburring operation the part must be presented to the cutting edge with high accuracy and proper registration whatever the depth of superfluous material has to be cut away. The process is quite similar to the fettling of castings (Chapter 18) except that the amounts of material to be removed tend to be far less so that the power requirements of the grinding or cutting tool are significantly reduced. For example, deburring is often carried out on small cast or molded parts, both metal and plastic, when for some reason a trim press is not applicable or is not capable of entirely removing the unwanted material. In considering the possibilities of automating or mechanizing the process of deburring, it is of importance to note that very different techniques will be needed for different materials. Plastics are much softer than brass, which is in turn less brittle than cast-iron. Thus the choice of tool can be as important as the technique adopted. Confronted with such an array of variables it is germane to ask whether a robot with its versatility and ability to change programs rapidly might be a suitable device for the deburring role. The answer is in the affirmative, although it must not be denied that the application is a difficult one which makes special demands on both the robot and associated equipment.

Basically a robot used in deburring will need to execute somewhat complicated motions at speeds which must be very closely controlled. In most cases the part, if the robot is to grasp and hold it, should not be too heavy and not beyond the ability of the robot to manipulate it. Conversely, if the part is held steady while the tool is grasped and moved around by the robot, then the tool, unlike the case of fettling, will not be heavy since much smaller amounts of material need to be removed.

Robot requirements for deburring

Specifically, the requirements of a robot in the deburring application will be the following.

1: CONTOURING CAPABILITY
The ability of the robot to follow a contour, however complex, is a necessity, and the path accuracy must be high if tolerances consistent with those demanded from normal machine-tool fabrications are to be realized.

2: REPEATABILITY
Robots may be noted for their repeatability, but since most small parts requiring to be deburred are of complex shapes, the robot must prove to be consistent in its path-following, however many twists and turns it has to make per operation. This imposes certain restraints on the robot designer since the robot must move along its programmed path without noticeable vibrations, for these would appear as variations in the final dimensions of the part. Any deviation from the path in excess of a few tenths of a millimeter will usually be unacceptable.

3: SPEED
It is unfortunate that in typical deburring operations several thousand similar parts have to be operated on individually. In the interests of high productivity, therefore, speed is of the essence. However, since the range of size of parts to be deburred can vary widely, a considerable

range of controlled speeds is desirable in a robot to be used in a deburring shop producing a multiplicity of parts. Further, the ability to vary the speed at which the robot describes the contoured path, moving faster over some parts than others, is essential if production rates are to be optimized.

4: PROGRAMMING

Because the robot will be called upon to deburr a wide range of parts, the ease with which the robot can be programmed and the speed at which it can change programs will be important, especially if short runs are envisioned. With this in mind, features such as the ability to edit and change parts of the program with the aim of optimizing the path will be a distinct advantage, and this applies not only to the path itself, but also to the speed at which the robot describes the path.

The above characteristics are possessed only by the most sophisticated robots. The deburring application is relatively new and not yet widely adopted, but the advent of the microprocessor will undoubtedly influence the use of the robot for this task.

It is not only the robot which requires sophistication in the deburring role. Associated tools must often be designed for specific purposes such as reaching into inaccessible regions of the part. Tools based on reciprocating actions rather than rotating parts are sometimes the only solution. The shape of the cutting edge deployed against the part is determined largely by the job to be done; cleaning up a thread will require a very different tool than would be needed to deburr a blind hole.

To achieve the accuracies already commented upon the support for the tool must be rigid. If the part can be presented accurately to the robot then whether the robot holds the tool or the part, the deburring operation will not be too complicated. This over-simplified statement, however, assumes that the part is presented to the cutting edge with appropriate accuracy over the entire programmed path, however complex. While some parts will be easy to work on, others will tax the ingenuity of the robot designer and operator alike. Nevertheless, robot deburring is probably the only available means of automating the process unless exceptionally long runs are scheduled, which would justify building special-purpose automation equipment.

Figures 21.1 and 21.2 illustrate typical robot deburring operations using ASEA robots.

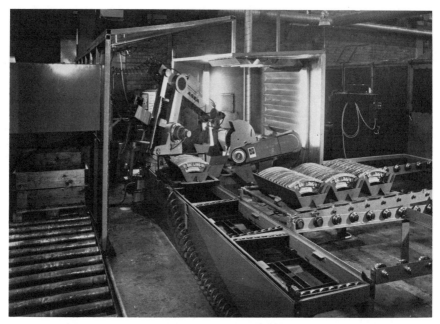

Figure 21.1 *Robot deburring operation at Kohlswa Steelworks, Sweden.*
By courtesy of ASEA

Figure 21.2 *Further example of robot deburring.*
By courtesy of ASEA

Chapter 22

Palletizing applications

In the manufacturing industries, stacking finished parts on pallets as they are fabricated is a means of automatically making up a load for removal to the next stage of manufacture or to the shipping bay or warehouse. It is clearly easier to pick up a single pallet loaded with parts than it would be to pick each part up separately, especially when the part needs to be properly orientated in order for the pick up device to grasp it.

In the distributive trades, a customer's order is frequently assembled on a pallet, the load being a selection of several different catalog items, each picked up from a different storage point. For such applications, highly developed automated devices have been introduced which can be programmed to move around a warehouse and select the required items from the shelves to make up the complete load ready for despatch. There are also special purpose palletizing machines available for stylized roles such as the packaging of candles, beer and soft drinks. In the brick-making applications described in Chapter 23, the stacking of bricks into a 'bench' is a further example of palletization, while the same process is to be found in the rubber and other industries.

This book describes numerous situations in which robots demonstrate their ability to pick up and set down articles with speed and precision. Robots have therefore a significant role to play in palletization of all kinds. The robot really comes into its own when the palletization pattern or make-up must change frequently, since by the selection of an appropriate program, the robot becomes equipped to deal with each new and different situation.

Robot use to achieve optimal pallet loading

The most interesting applications of robots to palletization demonstrate, therefore, considerable complexity in the program of events, with variety of movement being nicely combined with delicacy of grasp and positioning.

The requirement for palletization and depalletization cuts across many different industrial processes. Figure 22.1 illustrates a robot stacking plastic parts after molding (see Chapter 17). The following case

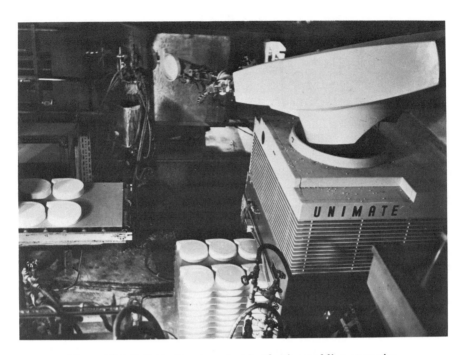

Figure 22.1 *Palletization by robot in plastics molding operation*

study takes a typical inventory situation where optimization of pallet loads is required.

A manufacturer's product is boxed in cartons of no less than fifteen sizes, each having its own unique weight. These boxes are conveyed to the warehouse for storing or shipping in pre-scheduled patterns. In each pattern, all boxes are of the same size and weight, but the number of boxes in the pattern may vary from 5 to 500. On average 50 to 100 boxes of the same type arrive to be dealt with.

The company has adopted a standard single faced pallet of size 40 inches by 48 inches. Different box sizes are never present on the same pallet load, but some pallets are only partially loaded when transferred into inventory. It is important not only to load the pallets properly but also to record accurately the contents and location of every individual pallet.

The following procedures may have to be carried out by the robot which is operated in conjunction with a programmable controller.

1 An empty pallet arrives at the loading station and the robot is required to load it to capacity with similar boxes. This results in a 'full' pallet.

2 An empty pallet arrives and is loaded to a prescribed pattern which does not fill the pallet to capacity. This results in a 'partial' pallet.

3 A partial pallet is brought from inventory and further boxes of the same type as those already loaded are added to it. Eventually this pallet will become a 'full' pallet.

4 A partial pallet arrives from inventory and further boxes are added but not in sufficient number to fill it to capacity, so it returns to inventory as a partial pallet.

This illustrates not only the complexity of robot operations necessary but also the recording requirements in up-dating the inventory as loads are varied. To this end every pallet must be identified unambiguously, this being accomplished by a bar-code label affixed to the side of the pallet and read by a pen-reader.

In this particular application, the pallet size has been carefully chosen to accommodate the maximum number of boxes of all types in the product range. Figure 22.2 illustrates the way in which boxes are set out in the pattern, some standing on one face, others on a different one, while the order is reversed at each level of the stack to provide a degree of interlock and therefore greater stability of the whole stack during its transportation into and out of its warehouse location.

The result is a complex program for the robot which must be capable of recognizing a wide range of possible situations and to determine, through its programming, just where it should move to pick up or set down its box correctly. A special hand, using vacuum pick-up techniques, has been developed specifically for this application. This

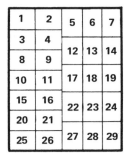

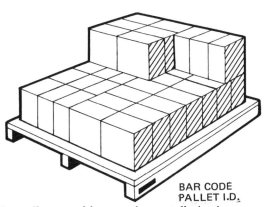

BAR CODE
PALLET I.D.

Figure 22.2 *Method of loading pallets to achieve maximum palletization*

was not an easy task since the weights of individual boxes varies between 16 and 53 pounds, so that a suction suitable for the heavier weights may be sufficient to damage the smaller boxes. Furthermore, due to the way in which the boxes are stacked in different ways to permit the pallet to be fully loaded, the hand must be able to approach boxes of widely different sizes from almost any aspect yet be able to grasp them firmly and set them down in the required position.

This sort of operation is to be found in many manufacturing operations in areas where finished or partially finished products arrive at a point where certain decisions have to be taken about the routing of the parts to the next stage of the operation. The robot can be linked to a programmable controller, permitting the supervisor to select a given program by punching a few buttons, after which the robot will carry out all of the required steps, thereby saving much labor. The robot knows what to do on receipt of certain instructions, but at the present state of the art it cannot itself recognize a 'situation' which confronts it and select its own program to deal with that situation. Such 'situations', presently beyond the intellectual capability of the robot, will include requirements for decision based on batch or routing changes, where on the one hand visual or tactile sense is essential, or where supervisory or managerial intervention may be called for.

Depalletizing by robot

In a processing plant, a pallet loaded with up to 40 paper sacks filled with raw material is presented to the robot. At this point the sacks are not located with any precision on the pallet, but are nevertheless crudely positioned simply because they are arranged in layers, in this case 8 layers each containing 5 sacks. The sacks weigh approximately 25 kilograms each, and they are picked up one at a time by a special vacuum gripper affixed to the arm of the robot. In order to introduce rather more precision into the operation, the robot then places each sack on to a locator fixture and then picks it up again, this time with the proper orientation for it to be presented to a slitting-emptying device which opens the sacks and allows their contents to be dropped at the required point. Because the number of sacks required for each mix in the process may vary, the operator controlling the process must be given a set of controls which will permit the sack number to be selected both at the start and the finish of the sequence. The operation can be further complicated by the following factors.

1 The incoming pallet may not be fully loaded with 40 sacks when it arrives at the robot pick-up position.

2 At the end of the load, there may not be sufficient sacks of raw material available to complete the mix. The robot must now

await the arrival of a new pallet load, and take from it the required number of sacks. This may result in a discrepancy in the number of sacks required for the next mix, so the problem may be compounded as the process continues since each new pallet is limited to a maximum of 40 sacks.

To meet these challenging demands the robot is provided with off-line programming, specifically a 'moving sequence control' (see Chapter 4) combined with a special operator-panel which allows the selection of the full range of possible 'Sack Start' and 'Sack Finish' conditions.

As in the brick-making industry, pallets can sometimes be extremely heavy when fully loaded, so generally it is the bigger robot which will be found in such applications. Similarly the complexity of some palletization and depalletization programs demands the use of a quite sophisticated machine. But the task is one to which the robot capability can be readily applied, so that palletization and depalletization are becoming increasingly important areas for these machines.

Chapter 23

Applications in brick manufacture

Bricks are in widespread use as a building component, and their manufacture dates back many years. The basic methods of manufacture have changed very little over the years, though automation has been introduced wherever possible, helped by the fact that bricks are required in vast numbers but the size and shape has remained virtually constant while other technologies have changed.

There are in fact two types of brick to consider in a section devoted to robot applications in the industry. Apart from the building brick which comes in various styles and finishes there is the refractory brick which is used to line furnaces and fireplaces. This class of brick must be capable of withstanding high temperatures without softening or cracking, and this requires them to be made by a process somewhat different from the conventional building brick.

Early bricks were made from mud and straw, hand-formed and dried in the sun. The advent of pottery techniques resulted in bricks being made from clay which was then fired in a kiln, and this is basically the process today. Mechanization in the industry did not start to occur until the nineteenth century. Prior to that period, bricks were made by placing molded and partially dried bricks in kilns and firing them for an appropriate time, after which the fire was allowed to go out and the kiln opened in order to cool. The process was time-consuming, but in the mid-1800's Hoffman introduced the continuous kiln technique. A circular kiln contained a dozen or more separate chambers. The principle was to direct the fire from chamber to chamber following a sequence which allowed bricks to be loaded, fired, cooled and removed on a continuous basis. Present day continuous kilns usually take the form of two parallel sections connected at the ends rather than being circular.

The brick manufacture process

The first stage in manufacture is to obtain the clay, which is then ground and crushed to a suitable grain-size and mixed with water to the appropriate consistency. To shape the bricks from this base material two methods are in use. The first, molding, also known as pressing, is to force the clay mix into a mold. Due to the very large volume of bricks required the mold is of course a multiple one, and the loaded mold will be heavy to move around.

The second process is one of extrusion. The clay is extruded into continuous columns which are then wire-cut into correct brick lengths in a manner similar to the method used to cut cheese. It is sometimes necessary to press extruded bricks to improve their shape and texture before they are fired.

Bricks made by either process are partially dried by being stacked in a cross-hatched pattern to permit air circulation. Today this is done on a kiln car, a tracked vehicle which will permit the brick stack to be moved to, first, a pre-drying area and thence to the kiln for firing. Much of this work has to be done by hand and is known as hacking. This is probably the most important step from the point of view of final product quality, yet it is a most unattractive job which damages the hands of the operatives, is boring, repetitive and tiring. The environment is also not a pleasant one in which to work. These seem to be classic reasons for using robots for the task, but the application is not an easy one for the robot to accomplish. The hacking process is very time-consuming, and this step in the entire brick-making process tends to set the overall production rate.

To complete the picture, however, the firing takes place as the cars loaded with bricks move through long kilns which are maintained at temperatures between 950 and 1200 degrees Centigrade, depending on the type of clay used and the nature of the final product. Firing time may be as long as four to ten days, so to maintain an effective production rate, the kilns must be very long, typically 400 feet. Unloading the kiln-cars is another operation usually performed by manual labor. In a typical case, eight men standing on a narrow platform above the kiln-car unload it by hand. When the car is empty it moves around an indexed track, pulled by a chain drive. At this stage of the process, broken or defective bricks must be removed and placed in a scrap truck or container, this requiring visual capability and judgement on the part of the operator — further reasons why the robot finds this a difficult job to take over.

Finally the bricks must be strapped or made up into stacks which can be shipped out. Protection of the corners of the stack (usually 100 bricks) must be provided by positioning cardboard or similar materials shaped for the purpose, and the stack banded with steel strip. Fork-lift trucks then remove five or six stacks at a time to the shipping area.

Despite this labor-intensive operation, production rates of 250 bricks per minute are achieved which are enough to fill about 16 kiln cars per day in a typical plant.

The robot contribution to brickmaking

Some early attempts were made to use robots in the brickworks, notably at the Burns Brick Company in Georgia, USA. It was hoped to reduce the manpower requirements to only three operators for each of the two shifts worked per day. The attempt was not simply to introduce robots, but also to use up-to-date transfer machinery wherever this proved feasible. Such machinery, it should be noted, is capable of working at rates almost twice as fast as those which human operators can attain. Hacking, however, resisted automation; it has not yet been possible to make a robot or any other form of automation work as fast and as reliably as a human operator.

The problems of the brickworks are in many ways related to palletization (see Chapter 22), a field where the robot is undoubtedly able to play a role. It is of interest to describe briefly some of the work carried out at the Burns Brick Company, since it provides a valuable insight into the whole problem of robotization of this industry. Since a brick does not acquire its full strength until it has been fired, it is essential that any handling equipment should be capable of taking up the bricks without damaging them, and this applies particularly to the hacking process which occurs at the pre-firing stage.

In an attempt to meet this requirement, Burns used two Versatran robots located on a platform straddling the kiln cars. The first robots were equipped with pneumatically operated 'bladder' hands capable of picking up five bricks at a time. These were to be set down in eight courses each containing six groups of a five-brick pick-up, or 240 bricks in all. The production rate achieved by human operators however was a difficult one to match, and to date it has not been possible to use a robot to the exclusion of manual labor in the brickmaking process.

There is one sector of the brickmaking industry however which is admirably suited to the robot. It is the manufacture of tar-bonded refractory bricks which are used for furnace linings and similar high-temperature environmental requirements. A standard form of this brick as produced by Dresser Industries in the USA is made from a mixture of magnesite and hot pitch and can be as much as 30 inches in length and weigh 80 pounds. Such bricks are formed in a mechanical press, and as each brick is made, it is removed from the press by an automatic picker which places it on a metal pallet. The pallet and brick then have to be picked up and loaded on to a kiln-car for transporting it into the curing oven. The brick emerges from the mold in a hot, relatively soft condition, and before the advent of the robot, laborers were required to fill the kiln-cars with pallets of brick. Since the combined weight of the pallet and its load was as much as 70 pounds, the job was both strenuous and difficult. Spurred by the success of the industrial robot in handling and positioning comparable loads in the automotive industries, Dresser Industries sought to employ them for this step in their own process. It transpired that the application was by no means as simple as it might at first sight appear, but the robot has proved equal to the task and is now being used very successfully.

Introduced in 1975, robots in this plant service three presses. It is common for two of the three presses to be in production at any time, the third being down for maintenance or for mold-changing (i.e. a program change for the robot, though this is accomplished very rapidly). One man monitors the entire operation whereas previously two men were required to service each press.

The pallet mentioned earlier is used in order to provide accurate registration for the robot pick-up. The automatic picker places the hot brick on to the pallet, and when three pallets are ready for transfer to the curing ovens the robot picks up all three simultaneously using specially-designed tong-like grippers. The pallets are slid on to shelves in the transport car. This is where the robot is on familiar ground (a palletization application) for when one shelf in the car is filled, the next pick-up of three pallets is placed on the next shelf. The car moves along an indexed line as it is fully loaded.

The best arrangement for the robotized line was found to be rather more complicated than that just outlined. The cars were divided into two halves, each of which had twelve shelves for holding six pallets at each level. Since the robot picks up three pallets at a time, it is required to go to each level twice. In the first operation, one half of the car is empty while the other half contains empty pallets. The robot picks up three filled pallets from an output conveyor and places them into the proper positions (as determined by the program) in the empty half of the car. It then goes to the other half of the car and picks up three empty pallets which it then deposits on to an input conveyor. The robot then picks up a further three full pallets and loads them behind the first three already placed in the car. This fills the shelf, so the next load is deposited at the next (lower) level. This sequence continues until one half of the car is loaded with full pallets leaving the other half completely empty. The robot then proceeds to perform the second operation as follows. The car indexing system shuttles the half-full car and other car loaded with only empty pallets into position. The robot loads full pallets into the second half of the first car, while removing empty pallets from the first half of the second car. When the second operation is completed the robot programs back to the first operation again while the fully loaded car is indexed into position for removal. Figure 23.1 illustrates the plant layout for this sequence of operations. This apparently complex series of operations minimizes the number of motions required of the robot and optimizes production rates.

The tooling required consists of arms on which the pallets can rest.

For placing pallets on input or output conveyors the pallets are deposited directly by the arms, but for loading into the shelves a pusher mechanism is required. The appropriate tooling is illustrated in Figure 23.2.

A final note on this application; the tar-bonded refractory brick is cured rather than fired during its manufacture. The final firing occurs after it has been installed in a furnace — that is, it achieves its final strength only after it has been placed in service. This means that it is vulnerable to damage at all stages of manufacture and transportation, and thus the process is designed to minimize the handling of the product. The robot never grasps the brick, only the pallet on to which it is placed by the automatic picker.

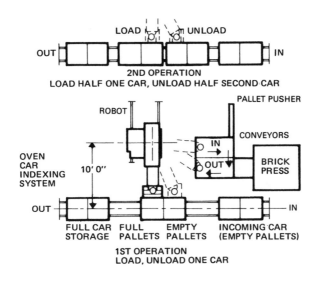

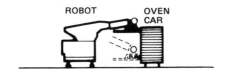

Figure 23.1 *Plant layout for robotized pallet handling in brick manufacture*

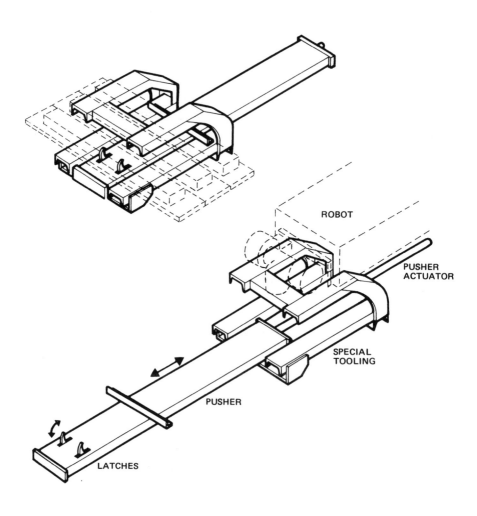

Figure 23.2 *Pusher mechanism for robot arms for placing pallets*

Chapter 24

Applications in glass manufacture

Glass consists of silica (sand), soda and lime, with small amounts of such chemical constituents as potash, lead oxide and boric oxide added to produce specific qualities in the finished product, for example to render it clear, colored or frosted. The ingredients have remained unchanged for centuries though the methods of production have changed considerably.

Outline of glass manufacturing process

Glass is made by fusing the materials at a high temperature and allowing them to mix together as they become molten. The result is a viscous mass of material which displays all the properties of a solid though it is technically a liquid in the supercooled phase. Very old glass windows in churches and similar buildings can be seen to be thicker at the base than at the top indicating that the material is still capable of flow under gravity, albeit at a very slow rate. The final product of the fusion process depends very much on the mix used. For example if glass is made simply by fusing sand, the result will be a rather crystalline form which will be very brittle. Adding soda and limestone improves the situation: chemically these additions reduce the melting point of the mixture from about 1700 degrees Centigrade to around 850 degrees, and improve the properties of the final product. The modern way of making sheet glass is the so-called 'float' process, introduced in 1959. Previously the fused mass of silica and additives had to be cast, rolled and finally polished to remove any distortions in the final surface. Somewhat surprisingly the float process did not evolve from previous experience but was based on entirely new technology. It changed the industry almost overnight.

The grinding and polishing of plate glass made by earlier methods resulted in up to 20% wastage and produced high capital and labor costs. In the float process, a wide ribbon of molten glass, anything up to about twelve feet in width, flows out of the furnace and floats on a bath of molten tin. The environment is controlled closely to permit any irregularities to melt out and allow the glass to assume the contours of the molten tin bed. This being flat, the glass settles into a completely flat sheet of even thickness if the ambient conditions are such as to allow the mass of material to cool slowly and form its natural shape under gravity. As the continuous glass sheet proceeds across the molten tin bed, it eventually becomes rigid enough to be picked up and run through rollers to anneal it without damaging the surfaces. The resultant product has a uniform thickness and needs no further finishing. There is, of course, a relationship between the density of the glass, the density of the tin and the rate of cooling which will determine the thickness to which the ribbon of molten glass will settle at as it cools. Fortunately this works out, in practice, as about one quarter of an inch, and this is the thickness which meets the demand of almost half the market for glass today. However technical improvements based on impeding or speeding up the flow of the ribbon have made possible the production of thinner and thicker glass sheets using the same float process. A thickness range of one tenth of an inch to no less than one inch is now possible with modern techniques and machinery.

Glass today comes in several forms. Plain glass is used for windows and mirrors, the larger units being of the thicker or plate glass forms. Patterned glass for decorative purposes is made by passing the sheet through suitably imprinted rollers while still in the semi-molten stage. Optical glass, used for lenses and prisms is specially chosen to be homogeneous throughout and free from discoloration. Safety glass, most important in automobile applications, is made from annealed glass which is subjected to further processes to toughen it and to produce properties which will cause it to shatter in small rounded bead-like particles when it finally breaks. Heat-resistant glass is made from a mix chosen to have a very low coefficient of expansion to minimize stresses in the glass when subjected to extreme temperature changes.

Generally speaking glass is a most difficult material to handle. It tends to be fragile, and sheets of glass will shatter if stressed, the resultant pieces being sharp and dangerous to handle. Serious accidents from glass making and handling are legion, yet the product is in widespread use in the home, the car and in the factory and office. Working glass is a craft. Glass blowers shape molten glass by blowing it, sometimes using molds. Flat sheets of glass are regularly cut to shape and edge-ground to meet a variety of needs, and the techniques involved, although they may look easy, demand a great deal of experience and the confidence which comes from regularly handling the material. Transporting and handling glass, particularly in large flat sheets is somewhat hazardous. Is there, then, any role for a robot to play in such a difficult arena? In certain applications, the answer is very much in the affirmative. It is of interest to discuss some of these applications which are typical of what can be achieved. Others will undoubtedly follow as experience is gained in the field as a whole.

Robot handling of sheet glass

Robots are usually at a disadvantage when compared with human beings because they lack dexterity and the freedom of movement provided by the human arms, hands and fingers. In handling glass, however, the robot can be provided with tooling which in some respects makes it superior to the human operator. When a man picks up a large sheet of glass, he must hold it around the edges, protecting his hands from the sharpness, the stretch of his arms limiting the width of the sheet which he can carry single-handed. The robot does not use this technique at all, but can be equipped with vacuum cups which press against the surface of the sheet to adhere to it so tightly that the sheet can now be picked up and moved around. Releasing the vacuum reduces the adhesion so that the sheet may be set down in a required place. Variations on this form of gripper can provide two sets of cups mounted back to back so that two pieces may be held simultaneously and dropped at different points if this is what the program calls for. The use of such vacuum cups has opened up a whole range of glass handling applications for robots.

The robot has no role as yet in the float process but only finds applications handling glass after it has been produced. The handling application can be extended further. Setting up a glass manufacturing plant, or a line manufacturing products such as television tubes from glass stock, requires that a wide range of operating variables be harmonized to ensure trouble-free production runs. This is due in part to the properties of glass, which has to be worked at the right temperature in the most suitable environment. To achieve these optimum operating conditions may take a lengthy set-up time, requiring a special crew to come on duty several hours before production personnel, so that everything is working properly when the shift commences. By the same token, if the line should have to be shut down for any reason at all, the outcome could be very costly. It is not uncommon for the line to be left running at all times during the shift. If there should be any equipment failures, the glass manufactured may be allowed to smash on the shop floor rather than stop a process which might take hours to re-establish. So, should a robot fail while working on such a line, the cost of such a failure could be a very serious matter, especially if a human operator was not able to stand in and carry on the robot's job until it was repaired. This is a very important consideration in deciding how to use robots in this industry.

An example of a robot handling flat sheet glass is that of edge-grinding windows for automobiles. Robots are not good at cutting glass. Although they have no difficulty in scribing an appropriate line holding a glass cutter, the sharp tap given by an experienced glass worker which causes the sheet to break around the line requires judgement which the robot does not possess. Further the second tap, necessary when the operator notices that the glass has not broken cleanly, is beyond the ability of the robot at this stage of its development.

Fortunately, however, the application of edge-grinding of automotive windows is one admirably suited to the robot; it is an application combining robotics and specialized automation, the latter being justified by the very large production runs typical of the automotive industry. The layout of the equipment for this operation is shown in Figure 24.1. The robot serves two edge-grinders. The glass sheets are picked up from an input conveyor using the vacuum cup technique. Each edge-grinder is loaded in sequence, the relatively lengthy grinding operation allowing the robot plenty of time to go through its programmed steps. Using the double-gripper system previously mentioned, the robot deals with two pieces of glass at a time, one a raw piece to be loaded into the grinder which is ready to receive it and the other a finished piece to be deposited on the output stacker.

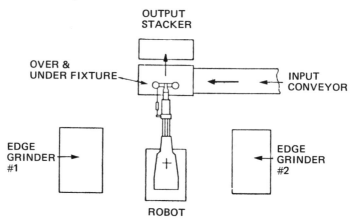

Figure 24.1 *Layout for robot in window edge-grinding operation*

The way in which the robot and specialized automation support one another in this role is illustrated by the fact that on arrival at the grinder an auxiliary unload-assist device comes into operation which raises the finished part to allow the robot's hand to move underneath the part into a nest on the grinder. Versatility is increased by providing the robot with independent programming of vacuum and/or blow-off on either set of vacuum cups. The hand is supported in such a way that the angular position of the vacuum cups can be varied plus or minus 20%, and the position of the cups relative to the wrist is adjustable. These features permit a wide range of part shapes to be handled. The robot can operate one or both of the grinders as required simply by selecting the appropriate program (Alternate Program Selection).

The continuous nature of the glass industry makes this application an exceedingly profitable investment since each robot replaces two workers per shift on a three-shift basis. The application is suited to windshields, side and back windows and any other flat or semi-flat glass shapes. In this application it is usual for the tool to be moved around the glass rather than have the robot manipulate the tool, this being made possible by the large production which justifies the use of special-purpose automation in conjunction with the robot. Figure 24.2 illustrates the special double hand used in this application.

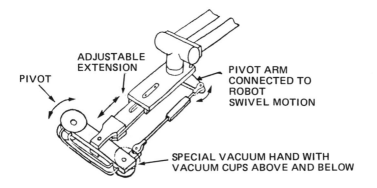

Figure 24.2 *Special double hand used in glass handling*

Glass in large flat sheets is conventionally transported in containers known as 'bucks'. The buck is a steel carrier inside which is a wooden crate containing the glass. This technique is used both for raw stock glass and for shaped parts such as windshields. The stack of glass inside the buck may be as thick as two feet, and due to its density, the weight of the whole container may be considerable. The buck is designed so that its entire face opens, displaying the stack of glass. Robots have been used successfully in unloading these containers. Using its vacuum cups, the robot flexes the top glass sheet slightly to reduce friction as it slides out the sheet.

Robot handling of fragile glass products

A more difficult application for the robot in glass manufacture is one which required it to move fluorescent tubes around in the factory where they were made. These tubes are made by pouring molten glass over a ceramic mandrel and allowing it to solidify. By the correct choice of the amount of molten material very thin tubes are formed which by their nature are very fragile and difficult to handle. Also they must be handled in the hot state since to allow them to cool before proceeding to the next step would result in very poor production rates.

In the late 1960's the problem was first attacked by equipping the robot with vacuum cups arranged in such a fashion that the robot could pick up 26 tubes per load. The configuration of the multiple vacuum cups was important since it was essential to brace these fragile tubes to some extent on pick-up to minimize breakage. It was found in this application that the hot 'new' glass is very abrasive, so that the vacuum cups quickly wore out and needed to be replaced. This led to further experiments which led to an entirely new concept for this application. Special grippers were designed for the robot embodying spring-loaded steel pins which provided some give on closure (see Figure 24.3) and the tubes were now picked up by their ends in a batch and moved to a pallet where they nested together. The robot program was based on 30-layer nesting in a total batch of 150 tubes. When the pallet was

Figure 24.3 *Robot lifts load of glass tubes*

fully loaded conventional transfer equipment moved it to the next manufacturing process. The conveyor system carrying the hot tubes was synchronized with the robot so that tubes were available to it with appropriate registration accuracy each time it moved in to make a pick-up. The system has proved quite successful and is still in use.

A sophisticated use of the robot in the glass industry is to be found in the manufacture of television picture tubes. These tubes are made in two main glass sections, the bulb or funnel-shaped body of the tube, and a panel, the flat plate which forms the front face of the tube on which the picture appears. The two pieces are cemented together using a heat-activated cement, and other processes are required in this manufacturing phase such as the addition of a conducting coating inside the bulb, the insertion of the electron gun and the positioning of a metallic shield perforated with hundreds of thousands of tiny holes. All of these steps are now robotized. Firstly the robot grasps the neck of the bulb and applies it to a coating machine which sprays and brushes on a coating of ferric oxide. Then the metal shield is inserted, followed by the electron gun. The panel is now placed in position. The whole operation is carried out in a furnace which serves to anneal the glass and to provide the heat required to activate the cement used. The assembly has now become a tube, but it has yet to be evacuated. From this point on the robot picks up the tube by its front panel and places it on an exhaust line where the tube is pumped out and finally sealed off. The tube is now capable of imploding so it must be handled with care, but the robot can accomplish this and places the finished tube on a transporter so that it can proceed to a test point prior to packing.

Most of the applications for robots in the glass industry do not tax the robot from the point of view of its positional accuracy and movement capability. The use of suction cups is particularly suited to the handling of glass, yet the fluorescent tube handling indicates that the grip of the robot can, when necessary, be made gentle enough so as not to damage the glass.

In addition to the applications which have been developed for the handling of glass sheet and glass artefacts by robots, some progress has also been made in the totally different field of glass blowing. The glass blower has for centuries shaped hollow bulb-like glass forms by heating and sealing the end of a glass tube and then blowing air into the open end while rotating the tube by hand in order to produce such vessels as bottles. The technique requires great craftsmanship to achieve the desired result. When several pieces of identical shape and size are required, the glass-blower will use an external mold which defines the final contours of his workpiece, though considerable skill is required to achieve a consistent thickness of wall.

The process of adding molten glass to a mold in just the right amount required to produce a part is analogous to this. In mass-production, a 'boule' of molten glass will be injected to a mold and then air introduced to force out the material so that it is shaped by being pressed against the faces of the mold. Some radomes have been made by this method, and robots have been used successfully to insert a rod into a mass of molten glass in a crucible, twist in order to gather up enough glass on the rod, and then feed the lump into a mold. The consistency of the robot's movements ensures a predictable charge of glass into the mold with consequent reduction of rejection rate.

Appendix:
List of principal
robot manufacturers

The following companies all offer one or more models of industrial robots. The listing is not exhaustive because only companies that are in series production and offer full application engineering and field service support were considered.

Almost daily one may hear of a new entrant into robot manufacturing or of an automation manufacturer who feels the urge to aggrandize his product with the label, ROBOT. Many have and surely many more will fall by the wayside. This listing may unjustly slight some responsible manufacturer, but at least it does no disservice to an innocent reader. These are responsible companies with a commitment to their respective robot product lines.

Europe

ASEA
    Electronics Division
    S-72183 Vasteras, Sweden

Comau Industriale
    Divisione Sistemi Saldatura
    Strada Orbassano n. 20/22-10095, Turin, Italy

Electrolux
    S-10545 Stockholm, Sweden

Hall Automation Ltd.
    (a GEC-Marconi Electronics Company)
    Colonial Way, Watford, Hertfordshire WD2 4FJ, England

Olivetti Controllo Numerico S.p.A.
    Strada Torino 603,10090 S. Bernardo d'Ivrea, Turin, Italy

Renault
    Machines — Outils
    Centre Parly 2, P.B. 70, 78150 Le Chesnay, France

Trallfa Nils Underhaug A/S
    Box 113, N-4341 Bryne, Norway

Unimation Europe Ltd.
    Stafford Park 4, Telford, Salop, England

VW — Volkswagenwerk AG
    3180 Wolfsburg, Germany

United States

Autoplace Inc.
1401 East Fourteen Mile Rd., Troy, Michigan 48084

Cincinnati Milacron
4701 Marburg Avenue, Cincinnati, Ohio 45029

Prab Conveyors, Inc.
5944 E. Kilgore Rd., Kalamazoo, Michigan 49003

Unimation Inc.
Shelter Rock Lane, Danbury, Connecticut 06810

Japan

Daini Seikosha Company, Ltd.
Industrial Electronics Division
31-1, 6-chome, Kameido, Koto-ku, Tokyo 136, Japan

Fujitsu Fanuc Ltd.
5-1 Asahigaoka 3-Chome, Hino-shi, Tokyo 191, Japan

Hitachi Ltd.
Nihon Bldg., 2-6-2 Ohtemachi, Chiyoda-ku, Tokyo, Japan

Kawasaki Heavy Industries, Ltd.
System Engineering Division
Kyoritsu Bldg. 22 1-3 chome, Shiba-koen, Minato-ku, Tokyo, 105, Japan

Mitsubishi Heavy Industries Ltd.
5-1, Marunouchi 2-chome, Chiyoda-ku, Tokyo, Japan

Nachi-Fujikoshi
World Trade Center, 2-4-1 Hamanatsucho, Minato-ku, Tokyo, Japan

Shin Meiwa Kogyo
5-9-29, Yako, Tsurumi-ku, Yokohama, Japan

Tokyo Shibaura Electric Co., Ltd.
1-6, Uchisaiwaicho 1-chome, Chiyoda-ku, Tokyo 100, Japan

Yasakawa and Co., Ltd.
Sen-i Boeki Kaikan
16-9, Uchikanda 2-Chome, Chiyoda-ku, Tokyo, Japan

Bibliography

References

Estes, Vernon E., *An Organized Approach to Implementing Robots*. November 1978

Markham, Edwin, *The Man with the Hoe and Other Poems*, Doubleday and Co., Inc. N.Y., N.Y. 1912

Ruzic, Neil P., *The Automated Factory — A Dream Coming True?* Control Engineering, April 1978, p. 58

Taylor, Frederick Winslow, *Scientific Management; Comprising Shop Management, the Principles of Scientific Management, Testimony Before Special House Committee; with a Forward by Harlow S. Person.* Greenwood Press, Westport, Connecticut. 1973

Books

Asimov, Isaac. *I, Robot*, Doubleday & Co., Inc., N.Y., N.Y. 1950

Capek, Karel. *R.U.R.* (Rossum's Universal Robots), Doubleday, Page & Co., N.Y., N.Y. 1923 (A Play)

Lundstrom, G., Glemme, B. and Rooks, B.W. *Industrial Robots-Gripper Review*, International Fluidics Services Ltd., Bedford, England. 1977

Tanner, William R., ed. *Industrial Robots*, Society of Manufacturing Engineers, Dearborn, Michigan. 1979

Reports

Abraham, R.G., Stewart, R.J.S. and Shum, L.Y. *State-of-the-Art in Adaptable-Programmable Assembly Systems*, International Fluidics Services Ltd., Bedford, England. 1977

Bolles, R. and Paul, R. *The Use of Sensory Feedback in a Programmable Assembly System*, Stanford University AI Memo-220, Stanford, California. October 1973

Japan Industrial Robot Association. *Specifications of Industrial Robots in Japan 1979*, Tokyo, Japan. 1979

Minsky, M.L. *An Autonomous Manipulator System.* Project MAC Progress Report III, M.I.T., July 1966; and subsequent Project MAC, Artificial Intelligence Memo Series, Massachusetts Institute of Technology, Cambridge, Massachusetts.

Project Committee of Unmanned Manufacturing System Design. *MUM: Methodology for Unmanned Metal Working Factory-Basic System Design*, Mechanical Engineering Laboratory, Tokyo, Japan. 1974

Society of Manufacturing Engineers. SME Technical Reports and Papers (Papers and reports abstracted and indexed quarterly in Technical Digest) Society of Manufacturing Engineers, Dearborn, Michigan. Numerous specialized reports published.

Stanford Research Institute and Stanford Research Institute International. Ongoing reports on projects sponsored by National Science Foundation, *Exploratory Research in Advanced Automation* and *Machine Intelligence Research Applied to Industrial Automation.* 1973

U.S. Air Force Materials Laboratory. ICAM (Integrated Computer-Aided Manufacturing) Reports. An ongoing program of the U.S. Air Force Materials Laboratory with ongoing reports.
> *Robotic System for Aerospace Batch Manufacturing.*
> > Task A: Sheet Metal Center — Robotic Drilling and Routing
> > Task B: Robotics Computer Software Design
> > Task C: Robot Assembly Capability

Robotic Support — National Bureau of Standards. Development of specifications and standards for software related to robots and their control systems.
ICAM Program Office, AFM/LTC Wright Patterson Air Force Base, Ohio

Yonemoto, Kanji. *The Present Status and the Future Outlook of Industrial Robot Utilization in Japan,* Japan Industrial Robot Association, Tokyo, Japan. 1978

In addition to the material cited, the Japan Industrial Robot Association publishes a wide variety of reports on industrial robots.

The National Bureau of Standards, a department of the U.S. Department of Commerce, is active in the field of robotics and regularly publishes reports. U.S. Department of Commerce/National Bureau of Standards, Washington, D.C.

Articles from periodicals

Hitachi Review, *Hitachi Hand-eye System,* Vol.22, p.362-365, 1977

Merchant, M.E., *The Future of Batch Manufacture,* Phil. Trans. R. Soc. Lond., Vol. 275A, p.357-372, 1976

Nevins, James L. and Whitney, Daniel E., *Computer-controlled Assembly,* Scientific American, Vol. 238, No. 2, p.62-74, February 1978

Rosen, Charles A. and Nitzan, David, *Developments in Programmable Automation,* Manufacturing Engineering, Vol. 75, p.26-30, September 1975

Will, P.M. and Grossman, D.D., *An Experimental System for Computer-controlled Assembly,* IEEE Trans. Comput., Vol. C-24, p.879-888, September 1975

Audio-visual material

Robot Video Tape Series
> *The Industrial Robot . . . An Introduction*
> *Industrial Robot Applications*

16 mm Films
> *Command Performance*
> *General Electric Manmate 2000*
> *Unimation Robots for People Who Think*
> *Computer Controlled Robot Assembly of Automobile Alternators*
> *Hierarchical Control (Computer Controlled Robot)*
> *Cincinnati Milacron 6CH Arm Robot*
> *Olivetti Company Robot*

Prab Robots in Industry
The DeVilbiss-Trallfa Robot
Society of Manufacturing Engineers and Robot Institute of America, Dearborn, Michigan

Periodicals

The Industrial Robot. Quarterly, International Fluidics Services, Ltd., Carlton, Bedford, England

Robot. Quarterly, Japan Industrial Robot Association, Tokyo, Japan

Robotics Age. Quarterly, Robotics Publishing Corp., La Canada, California

Robotics Today. Quarterly, Society of Manufacturing Engineers, Dearborn, Michigan

Proceedings of symposia and conferences

First National Symposium on Industrial Robots, Chicago, Illinois, USA. April 2 and 3, 1970. IIT Research Institute, Chicago, Illinois. 1970

Second International Symposium on Industrial Robots, Chicago, Illinois, USA. May 16, 17 and 18, 1972. IIT Research Institute, Chicago, Illinois. 1972

First Conference on Industrial Robot Technology, University of Nottingham, UK. March 27, 28 and 29, 1973. International Fluidics Services, Bedford, England. 1973

Second Conference on Industrial Robot Technology, University of Birmingham, UK. March 27, 28 and 29, 1974. International Fluidics Services, Bedford, England. 1974

Fourth International Symposium on Industrial Robots, Tokyo, Japan. November 19, 20 and 21, 1974. Japan Industrial Robot Association, Tokyo, Japan. 1974

Fifth International Symposium on Industrial Robots, Chicago, Illinois, USA. September 22, 23 and 24, 1975. IIT Research Institute, Chicago, Illinois. 1975

Third Conference on Industrial Robot Technology and Sixth International Symposium on Industrial Robots, University of Nottingham, UK., March 24, 25 and 26, 1976. International Fluidics Services, Bedford, England. 1976

Seventh International Symposium on Industrial Robots, Tokyo, Japan. October 19, 20 and 21, 1977. Japan Industrial Robot Association, Tokyo, Japan. 1977

Eighth International Symposium on Industrial Robots and Fourth International Conference on Industrial Robot Technology, Stuttgart, West Germany. May 30, 31 and June 1, 1978. International Fluidics Services, Bedford, England. 1978

Ninth International Symposium on Industrial Robots, Washington, D.C., USA, March 13, 14 and 15, 1979. Society of Manufacturing Engineers, Dearborn, Michigan. 1979

Tenth International Symposium on Industrial Robots and Fifth International Conference on Industrial Robot Technology, Milan, Italy. March 5, 6 and 7, 1980. International Fluidics Services, Bedford, England. 1980

Computer Vision and Sensor-Based Robots: Proceedings of a Symposium held at General Motors Research Laboratories, September 25 and 26, 1978. George C. Dodd and Lothar Rossol, eds. Plenum Press, N.Y., N.Y. 1979

Joseph F. Engelberger papers

Application of Robots in Die Casting. Technical Paper #35, Society of Die Casting Engineers. 1964

Role of Industrial Robots in Improving Production Operations. Automation, p.2-7, June 1964

Enter the Industrial Robot. Industrial Research, November 1969

Economic and Sociological Impact of Industrial Robots. Proceedings of First International Symposium on Industrial Robots, p.7-12, April 1970

Industrial Robots — Second Generation. Proceedings of Second International Symposium on Industrial Robots, p.211-233, May 1972

Robotics — Key Factor in Discontinuous Process Automation. Presented at 12th International Automation and Instrumentation Conference, Milan, Italy, November 1972

Industrial Robots — Reliability and Serviceability. Presented at Conference on Robots, Munich, Germany, November 1972

Metal Forming and the Unimate. Proceedings of Second Conference on Industrial Robot Technology, p.E3-39-46, March 1974

Three Million Hours of Robot Field Experience. Industrial Robot, Vol. 1, No. 4, p.164-168, June 1974

Four Million Hours of Robot Field Experience. Proceedings of Fourth International Symposium on Industrial Robots, p.133-148, November 1974

Production Problems Solved by Robots. SME Technical Paper #MS74-167, 1974

Robots in Material Forming Processes. Presented at Society of Manufacturing Engineers American Forming Conference and Exposition, May 1974

Man Robot Symbiosis. Proceedings of Fourth International Symposium on Industrial Robots, p.149-162, November 1974

Robotics, the Last Decade and the Next Decade. Kybernetes, Vol. 4, p.9-13, 1975

Robotics-like it was, like it is, and like it will be. Presented at IEEE Intercon Conference, New York, April 1975

Robots for Australian Industry. Presented at Australian Engineering Society Conference, May 1975

Robotics — A Cohesive Force in Advanced Manufacturing Technology. Presented at Westinghouse Machine Tool Forum, 1975

Performance Evaluation of Industrial Robots. Proceedings of Sixth International Symposium on Industrial Robots, p.J4-53-J4-66, March 1976

That's no Robot, That's Just Automation. Presented at Milwaukee Symposium on Automatic Computation and Control, April 1976

The Artificial Appendage Game. Presented at Second World Congress of the International Federation on the Theory of Machines and Mechanisms, 1976

Robots Produce Superior Shell Molds. Presented at the Society of Investment Casting, Amsterdam, Holland, 1976

Industrial Robots. Presented at International Harvester Worldwide Plant Managers Conference, 1976

Robots Make Economic and Social Sense. Atlanta Economic Review, p.4-8, July, August 1977

Robots Thrive in Hot, Hazardous and Boring Jobs. Presented at Westinghouse Machine Tool Forum, 1976

Will Robots Invade Assembly Operations?. SME Technical Paper //AD76-632, 1976

Designing Robots for Industrial Environments. SME Technical Paper //MR76-600, 1976

Robotics and CAD/CAM. SME Technical Paper //MS77-771, 1977

The Technological Contributions to Productivity. Proceedings of National Science Foundation's Second Symposium on Research Applied to National Needs (RANN), Vol. 3, p.109-110, 1977

A Robotics Prognostication. Proceedings of Joint Automatic Control Conference, 1977

Robotics and Society. Presented at Fourth International Congress of Cybernetics and Systems, 1978

Stand-Alone vs. Distributed Robotics. Presented at General Motors Symposium on Computer Vision and Sensor Based Robots, September 1978

Robot Arms for Assembly. American Society of Mechanical Engineers Paper //78-WA/DSC-37, 1978

Robotics in 1984. Presented at Robots '79, Nottingham, England, March 1979

Smart Robots for Smart Machine Tools. Presented at Sixteenth Numerical Control Society Annual Meeting and Technical Conference, March 1979

People, Robots and Automation in Assembly. Presented at Manufacturing Engineering Colloquium, Berlin, W. Germany, June 1979

Robots and Automobiles: Applications, Economics, and the Future. SAE Technical Paper //800377, 1980. Society of Automotive Engineers, Inc., Warrendale, Pennsylvania

Robot bibliographies

Collection of Abstracts of Technical Literature. 1978 Japan Industrial Robot Association, Tokyo, Japan

Robots — A Bibliography with Abstracts. 1978 NTIS/PS-78/0026 National Technical Information Service, Springfield, Virginia, USA

Coles, L. Stephen, Categorical Bibliography of Literature in the Field of Robotics. 1975 Technical Note 88-3, Stanford Research Institute, Menlo Park, California, USA

Index

Accident prevention, 89, 112. *See also*
  Safety provisions
Accommodation of robots, 237
Actuators, 33-4
Adaptivity of robots, 245
Advanced technologies, 128
AIDA Company stamping presses, 204
AiResearch welding robots, 178
Alcohol-ammonia fumes, effect of, 77
Alternate Program Selection, 272
Analog control, 23
  servo systems, 23
Analysis of First UTD Installation Failures
  (Macri), 91
Annealing, 247-8
*An organized approach to implementing
  robots* (Estes), 95-6
Application(s), prospective, 134-8
  studies of industrial robots, 139-276
Arc welding, 171-9
  danger to operative, 173
  flame cutting, 179
  process, 171-4
  programming, 176-7
  robots in, 174-9
  sequence, 175
  use of welding gun, 174-5
Areas of cost exposure, 109-10
Arm(s) of robot, 22
  computer control, 33
  configurations, 30, 31
  coordinates, xvi
  geometry, 30
  for indexing assembly line, 137
  performance, 36, 37
Articles from periodicals, 280
Articulations with point-to-point control,
  26-7, 32, 33
ASEA, arc welder, 178
  robots, 164, 255
Asimov, Isaac, xiii, xv, 3
  Laws of Robotics, 89, 118, 125
Assembly, 134, 135-8
  investigators, 136
Attitudes to robots, 112-13
Audio-visual material, 280-1

Aus-forming, 248, 249
Austenite formation, 248, 249
Automatic assembly, programmable, 130
  inspection, 129-30
  pickers in brickmaking, 266
  Programmable Assembly System (APAS),
    136
  warehousing, 129
Automation, of die casting, 13
  in paint spraying, 209
  foundry, 226
Automobile industry, painting problems, 207,
    210
  robots in, 212-15
  specialized in glass handling, 272
Automotive paint spraying, 134
  spot welding, 163-4

Belcher, J. V., 63n.
Bird report on reliability, 83
Bladder hands, 265
Blow molding, 219
Books on robots, 279
Boothroyd, Professor, 136
Brick manufacture, applications, 263-8
  process, 263-4
  robot contribution to, 265-8
  stacking, 257, 261
British Robot Association (BRA), xvi
Burns Brick Company, robots in, 265

California Wheel Company, use of robots, 227
Capek, Karel, 3, 4, 7
Carbon arc method of welding, 171, 172
Carbonized-silica particles, effect of, 77
Carcinogenic pigment in paint work, 209
Carry, Champ, xv
Cassettes, use of in programming, 14
Caterpillar Tractor factory, robot operations,
    250
Cartesian coordinates, 30, 32, 33
Casting cycle, factor affecting, 143
Chain link manufacture, 193-4
  plant layout, 193
  special gripper, 193
Chaplin, Charlie, 9

Characteristics of robots, 4, 5
    compared with humans, 5, 6
    extended specification, 8
Checklist of economic factors, 101-4
Chrysler spray painting complex, 214
Chuck alignment, automatic sensing, 245
Cincinnati Milacron welders, 178
    press robots, 204
Circuitry in special purpose automatic
    controls, 14
    for unload, quench and trim, 150
Classification of robots, 19-30
Coat-A-Matic arc welders, 178
Cold-chamber die casting machine, 142, 143,
    145
Comau robots, 164
Comecon interest in robotics, 116
Commercially available robot qualities, 117
Communication linking between robots and
    computers, 7
    and N.C. machines, 7
Complex payback example, 106, 107
Compliance centre for mating parts, 126
Computer Aid Manufacture (CAM), 129
Computer(s), control, 238
    of robot arms, 33
    for design (CAD), 7
    directed appendage trajectories, 122-3
    interpretation of visual and tactile data,
    122
    numerical control (CNC), 130, 234, 238,
    244
Computer-aided design (CAD), 130
Computer-controlled order picking, 129
Condec Corporation, xv
Continuous Path (CP) painting robots, 210
Control(s), 7
    systems, 19
    hierarchy of, 131, 132
    memory unit, 22
    playbacks, 23
Controller, 22
Conveyor, control of, 60, 61
Cost(s) and benefits, 101-4
    exposure, areas of, 109-10
    savings in safety equipment, 15
    versus human labor, 9-12, 235, 251, 252
Cylindrical coordinates, 30, 32

Daimler Benz spot-welding robots, 163
Danly automation system, 204
Dawson, Bryan L., 68
Debilitating jobs, elimination by robots, 112,
    116
Debugging, 16
Deburring metal parts, 253-6
    operations, 253-4
    robot, 256
    robot requirements, 254-5
Depalletizing by robot, 244, 260

programming of, 261
Depreciation costs, 103
Design factors, 75
Designing for industrial environment, 78-82
Deterioration, see Reliability
DeVilbiss Company, 210
Devol, George C., xv
Die care by robot, 151
Die casting applications, 101-57
    factors affecting casting cycle, 143
    forging, 190-1
    further considerations, 155-7
    lubrication, 181
    outline of operations, 141-2
    by robots, 12-15, 145-57
Die lubricant, effect on robot, 77
Direct numerical control (DNC), 130, 133,
    234, 244
Direct current (DC) servomotor system, 34
Disciplines useful to robotics, 179
Disposal of part by robot, 145-7
Double hand, in glass handling, 273
    for small bath machine system, 241
Downtime, 82
    unscheduled, 34
Draper Laboratories, investigations, 136
Dresser Industries use of robots, 265-6
Drive, mechanisms, 19
    motor, power to, 22
    systems for robots, 33-5
Drop forging, 189-90
Dual gripper, capability, 244
    design, 67
Dynamic performance and accuracy, 35-40

Economic factors, checklist of costs and
    benefits, 101-4
    commentaries, 102, 103
    justification, 125
    See also Costs
Edge-grinding windows for automobiles,
    271-2
Effect of robots on employment, 115
Ejection in die casting, 141
Electric arc welding, see Arc welding
Electrical, harness manufacture, 134
    noise and interference, 76
Electrically powered drives, 34, 35
Electromechanical drives, 33
    switching, 20
Electro magnets, 51
    pickup for flat surfaces, 54, 55
    power for, 55
Electronic control in sequencing, 62
Electronic/electrical elements in Unimate
    2000, 84
    reliability and failure rates, 83, 84
Employment, effect of robots on, 115
End effectors, hands, grippers, pickups and
    tools, 41-58

Energy conserving musculature, 123
Engelberger, Joseph F., papers on robots, 282-3
Environmental factors in robot systems, 75-8
Equipment layout, 66
Escort memory, 130
Estes, Vernon E., 95n., 97
Evaluation of worker attitudes, 114
Exoskeletons, 8
Expansion bladder hand, 52
Explosion, risk of, 77
Extrusion, brick, 264
    molding, 217-18

Failure rates, 83, 84
Fault tracing, total self-diagnostic, 124-5
Fettling, robotized, 228-30
Fiat use of robot spot welders, 168-9
Finance, cost justification, 103
    See also Cost and Economics
Fingers, self-aligning, 45
    for grasping different size parts, 45
Fire or explosion, risk of, 77
Fire hazard, in drive system, 35
    in paint shops, 209
Flame cutting, 179
Flash, in die casting, 141, 150, 151
    removal of, 134
Float process for glass manufacture, 269
Fluorescent tube handling by robot, 273
Ford Motor Company, robot press shop, 199
    robot safety, 90-1
Forging, applications, 189-96
    processes, 189-91
    robots in, 193-6
    working environment, 192-3
Foundry practice, 225-31
    casting process, 225-7
    robots in, 227-31
Fragile glass products, robot handling, 273-6
Friction, in mechanical gripper, 43
    in vacuum system, 49
Future attributes of robots, 117-25

Garrett Corporation, 178
Gas Metal Arc Welding (GMAW), 172
General application in seminar outline, 94
General Electric Company, robot induction program, 95-7
General Motors, and assembly, 136, 137
    spot welding robots, 163
    spray painting research, 214
    Unimates at, 16
General purpose hand, 123-4
German interest in robotics, 116
Glass, forms of, 270
    manufacture, 269-76
        process, 269-70
    robot handling of sheet glass, 271-3
        fragile glass, 273-6

Grasping, components for, 41-58, 60
    force, calculation of, 43, 44
    methods, 41-2
Gripping requirements, 67
Group technology, 128-9

Hacking, in brick making, 264
    resists automation, 265
Hammer forging, 189-90
Hand, cam operated, 46
    chuck type, 48
    double, 48
    mechanism for injection molding, 221
    special for cartons, 47
        for glass tubes, 48
    standard, 45
    wide-opening, 46
    with inside and outside jaws, 46
        modular grip, 47
        moveable jaw, 47
Handling hot metal billets, 195-6
    cylinders in forging operations, 195-6
    soft goods, 135
Hanify, Dennis W., 63n.
Hardening steel, 248-9
Harrow disc manufacture, 251-2
Hazardous situations, 77-82
Hazards in the industrial environment, 76
Heat, and the industrial robot, 76
    treatment, applications, 247-52
        effect of, 77
        processes 247-9
        robots in, 249-52
Heating torch, 55
Heat-resistant glass, 270
Heisenberg Uncertainty Principle, 36
Heliarc welding process, 171-2
Hierarchy of control systems, 131-2
High reliability of robots, 244
Hot-chamber die-casting machine, 142
Household robot, 135
Human labor, economic alternative, 11
    when needed, 82
    operator in die casting, 143
        health hazards, 77
    replacement by welding robots, 174
    and robot characteristics, 5
    size robot, 137
        Unimate 500, 138
    workers, improvement in working
        conditions, 111
Hydraulic v. electric drives, 33
    robot arc welder, 177
    servo system, 34
        fluids, 35
        oil leaks in, 34

Industrial robot analysis — working place studies (Hanify and Belcher), 63n.
Industrial robot, companies, 277-8

Industrial robot *(continued)*
  economic viability, 11
  environmental hazards, 77, 78
  evolution of, 3-7
  reliability outlook, 85
  three benefits, 16
  uptime and downtime, 82-4
  for welding, 174
  See also Robots
Industrie Roboter (Warnecke and Schafft), xvi
Inert gas welding torch, 56
Infrared switches, 62
Inherent safety, 125
Injection molding, 219
  plant layout for robot hand mechanisms,
    221
Inserts, load and positioning, 151, 153, 154
Installation costs, 102-3
Institute for Production and Automation in
  Stuttgart, 136
Integrated Computer Aided Manufacturing
  (ICAM), 133
Interaction with other technologies, 128
Interlock switches for correction, 28
Interlocks and sequence control, 61-7
  analysis, 66-7
  signals, 66-7
International Harvester, plant robot
  operations in, 251-2
International Symposium on Industrial
  Robots (ISIR), xvi
Investment calculations, 107-8
Investment casting, applications, 181-8
  mold making by robot, 186-7
  process, 181-4
I. Robot (Asimov),

Japan Industrial Robot Association (JIRA),
  xvi
Japan, machine shop automation, 233-4
  robots, 238

Kawasaki arc welders, 178
  robots, 204
Kawasaki Heavy Industries, xvi
Kohlwa steelworks deburring by robot, 212,
  255
Kuka robots, 164

Labor, cost escalation, USA, 10-11
  displaced, savings from 103-4
  force, appreciation of robot, 113
    improvement of conditions, 111-12
    physical danger prevented, 112
  See also Worker
Ladle, 56
Ladling by robot, 227
Lathe loading with double-handed robot, 239
Limbs, control of, 23
Limit switches mechanically controlled, 61

Limited sequence robots, 19-23
  programming, 22
Line tracking, 68-74
Liquid sprays, gases and harmful particles,
  76-7
Local and library program acceptance, 244
Locomotive devices, 9
Lubrication by robot, 151

Machine shop, constitution, 233
  automation, 233-5
Machine tool loading, application, 233-45
  robot application, 235-45
    attributes, 243-5
Machining cycle, analysis, 65
  sequence, 64-5
Macri, Gennaro C., 91
Magnetic induction of metal sheets, 53
  pickups, 51-5
  points for selection, 53-4
Magnets, 51
  electro, 51
  permanent, 51
Maintenance, needs and economics, 85-9, 91
  and periodic overhaul costs, 103
'Man Friday' concept for handicapped person,
  135
Man-robot voice communication, 124
Martensite formation, 248
Massey Ferguson robots on machine line, 239,
  241
  supervisory controller, 241
Matching robots to work place, 59-74
Materials handling sequence, 64-5
Mean Time Between Failure (MTBF), 125
  for Unimate system, 82, 83, 84, 85, 86
Mean Time To Repair (MTTR), 82, 125
Mechanical grippers, 42, 44-8
  friction in, 43
Mechanical hydraulic failure rate, 83-4
Memory unit, 23, 25
Metal, inert gas welding technique, 172
  parts, deburring, 253-6
Microprocessors, impact on robot control,
  177
Microswitches, 62
Mobility, 123
  in work area, 244
Modern Times (Chaplin), 9
Mold(s), 226
  care of, 228
  for castings, 181, 182
  making by robot, 184-8
    basic programs for, 186-7
  multiple produced from pattern tree, 183
  processes, plastic, 217-20
Motor driven robot, 34
Moving-base line tracking, 68
*Moving line application with a computer
  controlled robot* (Dawson), 68

Multiple appendage hand-to-hand coordination, 122
Muscle power, xvi

National Bureau of Standards (NBS), 133, 136
Near relation to robot, 7-9
Noise in paint shop, danger of, 209
Numerically controlled (NC), 7, 130, 234, 241, 244
machine tools, 7
machines, 235

Obsolescence, resistance to, 17
Occupational Safety and Health Act (OSHA) regulations, 91
Operator compared to robot, 4-5
instructing playback robot, 27
Operating power costs, 103
Operational risks, identifying, 90-1

Package distribution, 134
Packaging, 134
Paint, research in, 207
behavior, 207-8
spraying robots benefits analysis, 210-12
future of, 212-15
See also Spray painting
Painting, technique, 207-8
Palletization, 60, 121
applications, 257-61
in brickmaking, 266
depalletizing, 260-1
robot operating, 243, 257-60
Parameters for return on investment example, 108
Paraplegics and robots, 135
Part orientation, 59-60
in plastic molding, 222
rules for, 61
Part positioning, 65
Parts, cleaning, 134
handling, 129
plastic, loading, 257-9
Pattern, for casting process, 181
trees, 182
multiple molds produced from, 183
Payback, 107, 108
analysis, 11-12
calculation, 104
evaluation of robot costs, 11-12
simple, 104-5
Peg boards, 22
Periodicals on robots, 281
Permanent magnets, 51-2
Photoelectric devices, 62
Physical danger, robot prevents, 112
Pick and place robots, 20
Pickups, magnetic, 51-5
tools for, 55-8
vacuum, 49-51

Plastic molding application, 217-23
processes, 217-20
robots in, 220-3
parts, method of loading, 259
palletization by robot, 257-8
Playback robots, with continuous path control, 29-30
with point-to-point control, 22-9
Pneumatic, drive system, 33
motors, 35
nut-runners, drills and impact wrenches, 57
Polar coordinates, 30, 32
Power for robots, 7
PRAB press robots, 204
Pratt & Whitney Aircraft Group, shell mold making, 187-8
Press-forging, 191, 197
Press-to-press transfer of automobile sheet-metal parts, 40
Press work, applications, 197-205
operations, 195-205
dangers of, 198
robots in, 199-205
safety precautions, 198
Pressure switches, 62
Priorities in attribute development, 125
Proceedings of symposia and conferences, 281
Production rate, impact on payback, 105-6
increase, 199
Productivity gains in farming, 115
Program, selection for cut-out or alternate action, 243
switching, 251
timing, 39, 40
Programmable controller with triple robot installation, 241
Programmable Universal Machine for Assembly (PUMA), 136, 138
Programmed Article Transfer (Devol), xv
Programming the robot, 176-7
changed by cassette, 14
Project appraisal by the payback method, 104
Prosthesis, 8, 135
Psychological aspects of robot implementation, 95n.
Purchase price, costs, 102
Purdue University investigation, 136
Pusher mechanism for robot arms in brickmaking, 267

Qualities sought for the future, 118
Quality improvement savings, 104
Quenching in die casting by robot, 60, 145-7, 149, 150

Reaction time, 16
Recognizing parts by vision, 120-1
tactile sensing, 121-2
References, 279

Reliability, field experience, 86
   long range outlook, 85
   maintenance and safety, 75-91
   targets, 82
   theoretical assessment, 83-4
Remote Centre Compliance (RCC), 127
Repetitive work, robots and, 6
Reports, 279-80
Reprogramming, 23
Return on investment (ROI), 108
   evaluation, 107-8
Revolute coordinates, 30
Rhode Island University assembly
   investigation, 136
Robot(s), advantages over human labor, 4-5
   anatomy, 18-40
   applications to industry, see Chapters in
      Part II and entries throughout index
   appreciated by workforce, 113
   attitude to, 107
   bibliographies, 283
   characteristics, 4, 5
      compared with humans, 5-6
   classification, 19-30
   cost and benefits, 101-4, 109
      v. human labor, 9-12, 235, 251, 252
   definition, 8
   designing for industrial environment,
      78-82
   economic alternative to human labor, 11,
      15
   economics of, 101-10, 222
   effect on employer, 115
   environmental hazards, 75-8
   future capabilities, 117-38
   industrial, 3-7, 16, 17
   labor saving, 202, 211, 222, 234
      displacement, 251, 273
   maintenance costs, 201, 252
   near relations, 7, 8
   organizations, xvi
   payback evaluation, 11
   physical danger, 89, 112
   population increase, 115
   resistance to obsolescence, 17
Robot Institute of America, xvi, 8
Robotics, xiii, xvii
   Asimov's Laws of, 125
   Comecon interest, 116, 120
   engineer in, 61
   first law of, 89
   international industry, xvii
   organizing to support, 93-9
   three laws of, 3
   workforce acceptance of, 97-9
Roll forging, 191
Rooks, Dr Brian, on fettling problem, 230-1
Rotation molding, 220
Rough trimming by robot, 147
Routers, sanders and grinders, 57

Rudimentary vision, 120-1, 245
R.U.R. (Rossum's Universal Robots) (Capek),
   3

Sack finish in palletization, 261
   start in palletization, 261
Safety glass manufacture, 270
Safety in hazardous situations, 77-8
   levels and precautions, 89-91
   provisions, 156-7
   regulations, robot fitted for, 201
Schafler, Norman I., xv
Scrap and reject rates, reduction in, 15
Seam welding by robot, 29
Self-diagnostic fault tracing, 124-5
Sensors and sequence control, 60, 61
Sensory perception evaluation, 127
Sequence control, 60-7
   problem, example, 63-7
   system, setting up, 63-7
Sequence in press shop, 202-3
Sequencing of stepping switch, 22
Service industries, 135
Servo systems, analog, 23
   direct current, 34
   hydraulic, 34
   single articulation, 37
Sheep shearing by robot, 135
Sheet fanner or separator, 53
Sheet glass manufacture, 269-70
   robot handling of, 271-3
Shell mold making, 187-8
Shock and vibration, effect of, 76
Signals from robot to machine, 67
Simple payback, 104-5
Simple vacuum cup hand, 52
Single articulation servo system, 37
Slag removal damage, 77
Small batch machining system, double hand
   for, 241
   integrated robot NC system for, 242
Sociological impact of robot, 111-16
Sparks in welding, effect on robot, 77
Spatial intrusion minimized, 123
Special-purpose automation v. robots, 15-17
Special tooling costs, 102
Specmanship, 37
Speed evaluation, 36
Spin-control hand for twirling investment
   molds, 185
Spot welders, robot, coordination of, 169-70
Spot welding, 158-70
   automobile bodies, 170
   guns for, 56, 161-3
   metals for, 161
   outline of operation, 159-63
   planning robot line, 164-70
   robots in, 163-4
   sequence, 160-1
Spray gun, 57, 208

Spray painting, 207-15
  automation in, 209-10
  dangers in, 209
  environment for, 208-9
  robots in, 210-15
  See also Paint spraying
Stand alone *v.* distributed robotics, 130
Stanford Research Institute (SRI)
  International investigation, 136
  sensory perception evaluation, 127
State-of-the-art technology, 118
Stationary-base line tracking, 70-4
  implementation, 71
Statistics on robot population, 115
Stop button emergency, 90
Stud-welding head, 55
Surface hardening, 249

Tactile sensing, 121-2
Tar-bonded refractory bricks, robots used for,
  265
Teach pendant, 27-8
  training process, 28
Technical Center Laboratories (General
  Motors), 214
Technology, future attributes, 118
Telecherics, 9, 19
Television tubes manufacture, robot use in,
  275
Tempering and hardening, 248-9
Thermoforming, 219
Thermoplastics, 217
Throughput increase, savings, 104
Tools, 55-8
  changing, 58
  fastened to robot wrists, 55-8
Tracking window, 72
Training process for playback robots, 28-30
Training system, manufacturer's example,
  93-7
Trallfa Niles Underhaug, arc welder, 178
  spray-painting robot, 210-12, 214
Transfer line, advantages, 241
  intention, 234
  machines, 237
Travelling to work robots, 73-4
Trim press operation, robots in, 13-14
Trimming, by robot, 149, 150
  steel castings, 228, 229
Tungsten Inert Gas Process (TIG), 172
Turbine blades manufacture, 187-8

Unimate robot, arc welding, 177
  die casting, 145
  in General Motors, 88
  press robots, 204
  systems, reliability estimate, 83-6
  track mounted, 73

Unimates, xv, 35, 38, 83, 84, 107, 136, 138,
  153, 154, 167, 168, 193, 201
Unimation Incorporated, xv, xvi, 17, 136
  outline on general application, 94
  for reliability, 86
  treatise of, 39
United States, Department of Defense
  viewpoint, 119
  job dissatisfaction, 10
  labor cost escalation, 10-11
Unloading by robot, 145-7
Unmanned factory concept, 128, 131
Unmanned manufacturing from robotics and
  numerical control, 245
Upset forging, 191

Vacuum, pickup systems, 52
  switches, 62
  systems, 49-51
    cup hand, 52
    cups, 49
    pickup systems, 52
    pump *v.* venturi, 50-1
VAL computer language, 137-8
Velocity traces for short and large arm
  motion, 37, 39
Venturi *v.* vacuum pump, 50-1
Versatran, robots in brickmaking, 265
  press robots, 204
Vibra rail in foundry work, 227
Vision, rudimentary, 120-1
Volkswagen, robots of, 204
Volvo spot welding installation, 166

Warnecke and Schrafft, work of, xvi
Welding, 60
  arc, 171-9
  guns, 161-3
    robot using, 29
  sequence, 160-1
  spot, 158-70
Westinghouse investigations, 136
Work configuration classified, 67
Work Force Acceptance Checklist, 97-9, 114
Work force, acceptance of robots, 97-9
  robots entering, 116
Worker attitudes, evaluation, 114
  See also Labor force
Working life, quality improvement, 111-12
Workpiece feed position, 65
Workplace layout, 67-74
Wrist articulations, 32

Xerox Corporation, robots of, 238